Pathogenesis Study of Glioma: From Glioma Stem Cells, Genomic Tags, to Rodent Models

Pathogenesis Study of Glioma: From Glioma Stem Cells, Genomic Tags, to Rodent Models

Editor

Hailiang Tang

MDPI • Basel • Beijing • Wuhan • Barcelona • Belgrade • Manchester • Tokyo • Cluj • Tianjin

Editor
Hailiang Tang
Department of Neurosurgery
Huashan Hospital, Fudan
University
Shanghai
China

Editorial Office
MDPI
St. Alban-Anlage 66
4052 Basel, Switzerland

This is a reprint of articles from the Special Issue published online in the open access journal *Brain Sciences* (ISSN 2076-3425) (available at: www.mdpi.com/journal/brainsci/special_issues/patho_glioma).

For citation purposes, cite each article independently as indicated on the article page online and as indicated below:

LastName, A.A.; LastName, B.B.; LastName, C.C. Article Title. *Journal Name* **Year**, *Volume Number*, Page Range.

ISBN 978-3-0365-6391-6 (Hbk)
ISBN 978-3-0365-6390-9 (PDF)

Contents

Editorial

Pathogenesis Study of Glioma: From Glioma Stem Cells, Genomic Tags, to Rodent Models

Hailiang Tang [1,2,3,4,5,†], Xi Li [6,†] and Rong Xie [1,2,3,4,5,*]

[1] Department of Neurosurgery, Huashan Hospital, Shanghai Medical College, Fudan University, No. 12 Wulumuqi Middle Road, Shanghai 200040, China
[2] National Center for Neurological Disorders, Shanghai 200040, China
[3] Shanghai Key laboratory of Brain Function and Restoration and Neural Regeneration, Shanghai 200040, China
[4] Neurosugical Institute, Fudan University, Shanghai 200040, China
[5] Shanghai Clinical Medical Center of Neurosurgery, Shanghai 200092, China
[6] Shanghai Pulmonary Hospital, Tongji University, Shanghai 116085, China
* Correspondence: rongxie@fudan.edu.cn
† These authors contribute equally to the paper.

Citation: Tang, H.; Li, X.; Xie, R. Pathogenesis Study of Glioma: From Glioma Stem Cells, Genomic Tags, to Rodent Models. *Brain Sci.* **2023**, *13*, 30. https://doi.org/10.3390/brainsci13010030

Received: 19 December 2022
Accepted: 21 December 2022
Published: 23 December 2022

Glioma remains the toughest brain tumor among all primary central nervous system (CNS) tumors. The complexity of its pathogenesis makes it difficult to achieve radical cure, especially in the case of glioblastoma multiforme (GBM, WHO grade IV), the most aggressive subtype of glioma [1]. Recently, the existence of glioma stem cells (GSCs) in GBM has been demonstrated, which exhibit the properties of neural stem cells (NSCs) and are responsible for chemo/radiotherapy resistance and tumor recurrence [2–4], making them potential therapeutic targets against GBM [5,6].

In this Special Issue, we collected 10 articles discussing the genomic alterations of glioma, its immune microenvironment, and multi-therapy for glioma. *Wu.* et al. screened necroptosis (NCPS)-related genes (CTSD, AP1S1, YWHAG, and IER3) to construct a prognostic model for GBM, paving the way for the use of new targets for the diagnosis and treatment of glioma. *Fang* et al. proved that the transient receptor potential (TRP) family genes are promising immunotherapeutic targets and potential clinical biomarkers for glioma. Besides genetic mutations, increasing evidence shows that the tumor microenvironment is also important for glioma development and resistance to therapy [7]. Within the GBM microenvironment, *Alice Giotta Lucifero* et al. found that the phosphatase and tensin homolog (PTEN)-related immune landscape mainly consists of T_{reg} and M2 macrophages, which repress the antitumor immune activation and are responsible for triggering glioma cell growth and invasion. *Li* et al. demonstrated that the tyrosine phosphatase receptor type N (PTPRN) could be an independent prognostic factor and correlates with tumor immune infiltration in low-grade glioma.

In the past decade, newly emerging therapeutic strategies such as tumor-treating fields (TTF) have been introduced in GBM patients [8]. *Yu* et al. suggested that irreversible electroporation possibly mediates glioma apoptosis via the upregulation of transcription factor AP-1 and Bim (Bcl2l11) expression. Programmed death protein 1 (PD-1) and programmed death-ligand 1 (PD-L1) play critical roles in tumor immune escape, and several immunotherapies, including PD-1/PD-L1 checkpoint inhibitors and chimeric antigen receptor-T cells (CAR-T), have already been applied in glioma therapy. *Yu* et al. explore the PD-1/PD-L1 protein expression in recurrent glioma and its paired primary tumor and reveal a tendency of increased PD-1/PD-L1 in recurrent glioma. However, immunotherapies proved therapy failures, which strongly indicates that beyond the T cell-based adaptive immunity, innate immunity might also be the key to regulating anti-tumor immunity in the glioma microenvironment. *Qi* et al. provided a cellular response to the interleukin-4 (IL-4)-related gene signature as an excellent immune biomarker of gliomas, which may be beneficial to develop novel immunotherapies for glioma.

Rodent models of glioma are still indispensable for understanding the basic principles of glioma development and tumor invasion [9]. Rodent models, including xenograft (Figure 1) and genetically engineered models, are used to study glioma development, reveal tumor progression, and test novel therapy strategies [10,11]. *Cintia Carla da Hora* et al. proposed patient-derived xenografts (PDX) or patient-derived GSC models in glioma research, which provide the possibility of studying glioma growth, treatment response, and survival outcome. In addition, *Hannes Becker* et al. conducted multilayered profiling of a platelet-derived growth factor B (PDGFB)-driven glioma mouse model and discovered radiological, histological, and metabolic features that are comparable to human high-grade glioma.

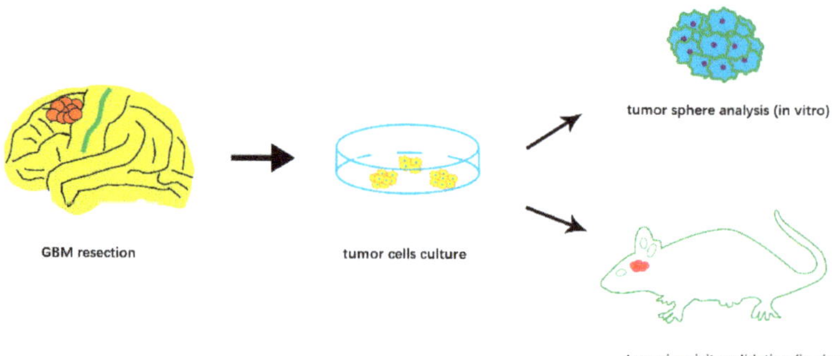

Figure 1. Schematic illustration of patient-derived xenografts (PDX) to study GBM biology.

Author Contributions: H.T. wrote the manuscript and drew the figure, X.L. collected the references, and R.X. revised the manuscript. All authors have read and agreed to the published version of the manuscript.

Institutional Review Board Statement: Not applicable.

Informed Consent Statement: Not applicable.

Data Availability Statement: Not applicable.

Conflicts of Interest: The authors declare no conflict of interest.

References

1. Matarredona, E.R.; Pastor, A.M. Neural Stem Cells of the Subventricular Zone as the Origin of Human Glioblastoma Stem Cells. Therapeutic Implications. *Front. Oncol.* **2019**, *9*, 779. [CrossRef] [PubMed]
2. Matarredona, E.R.; Zarco, N.; Castro, C.; Guerrero-Cazares, H. Editorial: Neural Stem Cells of the Subventricular Zone: From Neurogenesis to Glioblastoma Origin. *Front. Oncol.* **2021**, *11*, 750116. [CrossRef] [PubMed]
3. Lombard, A.; Digregorio, M.; Delcamp, C.; Rogister, B.; Piette, C.; Coppieters, N. The Subventricular Zone, a Hideout for Adult and Pediatric High-Grade Glioma Stem Cells. *Front. Oncol.* **2021**, *10*, 614930. [CrossRef] [PubMed]
4. Suvà, M.L.; Tirosh, I. The Glioma Stem Cell Model in the Era of Single-Cell Genomics. *Cancer Cell* **2020**, *37*, 630–636. [CrossRef]
5. Muftuoglu, Y.; Pajonk, F. Targeting Glioma Stem Cells. *Neurosurg. Clin. N. Am.* **2021**, *32*, 283–289. [CrossRef] [PubMed]
6. Piper, K.; DePledge, L.; Karsy, M.; Cobbs, C. Glioma Stem Cells as Immunotherapeutic Targets: Advancements and Challenges. *Front. Oncol.* **2021**, *11*, 615704. [CrossRef] [PubMed]
7. Barthel, L.; Hadamitzky, M.; Dammann, P.; Schedlowski, M.; Sure, U.; Thakur, B.K.; Hetze, S. Glioma: Molecular signature and crossroads with tumor microenvironment. *Cancer Metastasis Rev.* **2022**, *41*, 53–75. [CrossRef] [PubMed]
8. Kumaria, A. Observations on the anti-glioma potential of electrical fields: Is there a role for surgical neuromodulation? *Br. J. Neurosurg.* **2022**, *36*, 564–568. [CrossRef] [PubMed]
9. Hetze, S.; Sure, U.; Schedlowski, M.; Hadamitzky, M.; Barthel, L. Rodent Models to Analyze the Glioma Microenvironment. *ASN Neuro.* **2021**, *13*, 17590914211005074. [CrossRef] [PubMed]

10. Hicks, W.H.; Bird, C.E.; Traylor, J.I.; Shi, D.D.; El Ahmadieh, T.Y.; Richardson, T.E.; McBrayer, S.K.; Abdullah, K.G. Contemporary Mouse Models in Glioma Research. *Cells* **2021**, *10*, 712. [CrossRef] [PubMed]
11. da Hora, C.C.; Schweiger, M.W.; Wurdinger, T.; Tannous, B.A. Patient-Derived Glioma Models: From Patients to Dish to Animals. *Cells* **2019**, *8*, 1177. [CrossRef] [PubMed]

Article

Irreversible Electroporation Mediates Glioma Apoptosis via Upregulation of *AP-1* and *Bim*: Transcriptome Evidence

Shuangquan Yu [1,†], Lingchao Chen [1,†], Kun Song [1], Ting Shu [2], Zheng Fang [3], Lujia Ding [3], Jilong Liu [2], Lei Jiang [2], Guanqing Zhang [4], Bing Zhang [2,*] and Zhiyong Qin [1,*]

[1] Department of Neurosurgery, Huashan Hospital Shanghai Medical College, Fudan University, Shanghai 200040, China
[2] Intelligent Energy-Based Tumor Ablation Laboratory, School of Mechatronic Engineering and Automation, Shanghai University, Shanghai 200444, China
[3] Division of Biomedical Engineering, University of Saskatchewan, Saskatoon, SK S7N 5A9, Canada
[4] Biomedical Science and Technology Research Center, School of Mechatronic Engineering and Automation, Shanghai University, Shanghai 200444, China
* Correspondence: bingzhang84@shu.edu.cn (B.Z.); wisdomqin@vip.163.com (Z.Q.)
† These authors contributed equally to this work.

Abstract: The heat-sink effect and thermal damage of conventional thermal ablative technologies can be minimized by irreversible electroporation (IRE), which results in clear ablative boundaries and conservation of blood vessels, facilitating maximal safe surgical resection for glioblastoma. Although much comparative data about the death forms in IRE have been published, the comprehensive genetic regulatory mechanism for apoptosis, among other forms of regulatory cell death (RCD), remains elusive. We investigated the electric field intensity threshold for apoptosis/necrosis (YO-PRO-1/PI co-staining) of the U251 human malignant glioma cell line with stepwise increased uniform field intensity. Time course samples (0–6 h) of apoptosis induction and sham treatment were collected for transcriptome sequencing. Sequencing showed that transcription factor *AP-1* and its target gene *Bim* (Bcl2l11), related to the signaling pathway, played a major role in the apoptosis of glioma after IRE. The sequencing results were confirmed by qPCR and Western blot. We also found that the transcription changes also implicated three other forms of RCD: autophagy, necroptosis, and immunogenic cell death (ICD), in addition to apoptosis. These together imply that IRE possibly mediates apoptosis by the *AP-1-Bim* pathway, causes mixed RCD simultaneously, and has the potential to aid in the generation of a systemic antitumor immune response.

Keywords: irreversible electroporation; glioma; apoptosis; regulatory cell death; *AP-1*; *FOS*; *JUN*; *Bim*

Citation: Yu, S.; Chen, L.; Song, K.; Shu, T.; Fang, Z.; Ding, L.; Liu, J.; Jiang, L.; Zhang, G.; Zhang, B.; et al. Irreversible Electroporation Mediates Glioma Apoptosis via Upregulation of *AP-1* and *Bim*: Transcriptome Evidence. *Brain Sci.* **2022**, *12*, 1465. https://doi.org/10.3390/brainsci12111465

Academic Editors: Lucia Lisi and Terry Lichtor

Received: 21 September 2022
Accepted: 26 October 2022
Published: 29 October 2022

Publisher's Note: MDPI stays neutral with regard to jurisdictional claims in published maps and institutional affiliations.

1. Introduction

Glioblastoma multiform (GBM) accounts for the majority of gliomas (56%) [1] and displays the most aggressiveness. The efficacy of current surgery plus conventional radiation and chemotherapy for GBM is poor, despite a second surgical treatment [2,3]. GBM patients only have a median overall survival of 12–15 months [4] and inevitably die of tumor recurrence or progression. The development of novel treatment strategies is urgently required. As a novel non-thermal tumor ablation modality, irreversible electroporation (IRE) has successfully been applied to solid tumors, including liver cancer [5], pancreatic cancer [6], prostate cancer [7], and renal cell carcinoma [8], after Food and Drug Administration (FDA) approval in 2008. GBM is characterized by a rich blood supply and requires spatially maximal safe surgical resection with minimal disruption of neurological function. Compared to conventional thermal ablative technologies, IRE ablation would not be subject to heat-sink effects [9] from nearby blood flow and have clear ablation boundaries. Moreover, IRE effectively protects vascular networks within treated areas. These make it a suitable therapeutic candidate for GBM [10]. Within the field of GBM treatment, there was only modest evidence of animal research and in vitro cell experiments for IRE [11,12]. Further investigation into gene regulatory mechanisms and therapeutic efficacy after IRE must be completed.

Selectively based on specific signaling pathways, there have been numerous studies for gene regulatory mechanism of regulatory cell death (RCD), including apoptosis after IRE. Loss of mitochondrial membrane potential, cytochrome c release, caspase-9 activation, upregulation of pro-apoptotic factors (BAX, BAK, BAD), and downregulation of anti-apoptotic factors (Bcl-2, Bcl-xL, Mcl-1) confirmed the involvement of mitochondria, mostly through the intrinsic apoptosis pathway. However, in some studies, BID cleavage also points to the activation of type II extrinsic-like apoptosis. In some cells (HCT, B16F10, E4 SCC, Jurkat), apoptosis also progresses through the type I extrinsic-like pathway without or with little involvement of mitochondria and with caspase-8 activation and modulation of extrinsic apoptotic regulators, which influence sensitivity to nsEP [13]. To the best of our knowledge, few studies were based on high-throughput transcriptome sequencing in this gene regulatory mechanism issue after IRE in glioma. The transcriptome includes the complete set of transcripts in cells. Based on the generation of a high-throughput sequencing technology platform, transcriptome sequencing is used to comprehensively and quickly obtain almost all transcriptional sequence information of cells during a certain state. Unbiased high-throughput transcriptome sequencing seems to be the most valid approach to avoid any bias during pathways pre-selection.

Electric field strength threshold values for U251 cell apoptosis/necrosis were identified under stepwise increased field intensity. Cells of apoptosis induction by IRE and sham intervention were sampled (0–6 h) for high-throughput transcriptome sequencing. Furthermore, the obtained results disclosed mixed RCD forms simultaneously (including autophagy, necroptosis, and ICD) [14]. Here, we aimed to characterize the mechanisms of IRE-mediated RCD in vitro using the U251 cell line by high-throughput transcriptome sequencing. We then confirmed the sequencing results by qPCR and Western blot. We hypothesized that IRE-mediated RCD occurs via the AP-1–Bim-related signaling pathway. Mechanistic characterization of IRE-induced RCD will be important for translation into brain tumor clinical trials, in which IRE-mediated RCD will be exploited to reduce recurrence and inhibit the progression of GBM.

2. Materials and Methods

2.1. Cell Culture and IRE

U251 human malignant glioma cells were maintained in Dulbecco's modified eagle medium (DMEM) (hg)-10% FBS in 5% CO_2, 37 °C. Cells were seeded into 6-well plates before electroporation. Tumor cells were exposed to a flat electric field generated by a custom-made IRE pulse generator (HUIWEI MEDICAL Inc., Suzhou, China) after reaching 70–80% confluency, with field intensity incrementally increased. The pulse parameters were set as pulse duration 100 µs, frequency 1 Hz, 60 pulses, and unipolar. For DAPI staining, samples were incubated in 10% DAPI for 20 min before the intervention. Flat non-uniform/uniform electric fields were generated by two parallel-needle electrodes (blunt-ended, diameter: 0.4 mm, center-to-center distance: 1 cm, perpendicularly oriented to the plate bottoms, made with stainless-steel) or two parallel-plate electrodes (thickness: 0.4 mm, width: 2 cm, height: 4 cm, edge-to-edge distance: 1 cm, both vertically erected on the plate bottoms, made with stainless-steel) in 3 mL PBS at the center of every well of the 6-well plate. The pulse parameters were set as pulse duration 100 µs, frequency 1 Hz, 60 pulses, unipolar, and applied voltage of 1100 V on the needle electrodes or 200–730 V on the plate electrodes. Three-dimensional numerical simulations of the flat non-uniform/uniform electric fields were modeled using the finite element software (COMSOL Multiphysics 5.6, COMSOL Inc., Stockholm, Sweden, 2021).

2.2. Apoptosis and Necrosis Assay

Field strength threshold values for cell apoptosis/necrosis were identified by DNA fluorescence staining (YO-PRO-1/PI co-staining) and morphological changes 24 h post electroporation. YO-PRO-1 staining permits real-time cytofluorometric analysis of apoptosis with cell viability unaffected [15]. For YO-PRO-1/PI co-staining, samples were incubated in 5% YO-PRO-1/PI immediately following the electroporation. YO-PRO-1 and PI resulted in green and red staining of cell apoptosis and necrosis separately. Pictures were taken using a Nikon DS-Ri2 Zoom fluorescent microscope which can splice approximately 100 single fields of a normal microscope. Each picture showed the full view of a well in a 6-well cell culture cluster. Fluorescence intensity after flat uniform electric field treatment was analyzed using ImageJ software (ImageJ 1.46, the National Institutes of Health, Bethesda, MD, USA, https://imagej.nih.gov/ij/ (accessed on 1 July 2021)). S-shaped curves were obtained after all data were fitted by a nonlinear sigmoid curve regression using GraphPad Prism Software (Prism 7.00, GraphPad Software Inc., CA, USA, 2021, the x-axis displays the field intensity, and the y-axis shows the percentages of the fluorescent area in relation to the entire uniform electrical field-treated area) (https://www.graphpad.com/scientific-software/prism/ (accessed on 2 July 2021)). After electroporation, cells were re-cultured for another 24 h to further characterize their states by light microscopy.

2.3. RNA Extraction and Transcriptome Sequencing

Flat uniform electric fields of 520 V/cm were generated by the two parallel-plate electrodes in 3 mL PBS at the center of every well of the 6-well plate. Normal culture medium was added after IRE. Sham treatment was similarly handled in parallel with no current delivery. Samples were collected at 0 h after sham treatment, 0 h, 3 h, and 6 h after the electroporation. Cells exposed to the electric field were scraped off plates by a cell scraper and were harvested by centrifugation at 1000 rpm for 5 min. A total of 1 mL of TRIzol reagent was then added and mixed thoroughly by vortexing. Three replicates were taken for samples of each time point. All samples were stored at $-80°$ until total RNA extraction. Total RNA was extracted using the mirVana™ miRNA ISOlation Kit (Ambion-1561) according to the instructions of the kit. The purity and quantification of RNA were assessed using the NanoDrop 2000 spectrophotometer (Thermo Scientific, Waltham, MA, USA). RNA integrity was evaluated using the Agilent 2100 Bioanalyzer (Agilent Technologies, Santa Clara, CA, USA). Libraries were constructed using the TruSeq Stranded mRNA kit (Illumina, San Diego, CA, USA) according to the manufacturer's instructions. These libraries were sequenced on the Illumina transcriptome sequencing platform (NovaSeq 6000 Sequencing System).

2.4. Western Blot

Western Blot was performed using standard Western blot methods. c-Fos and Bim were detected using an Anti-c-Fos antibody (ab190289, Abcam, Cambridge, UK) and an Anti-Bim antibody (ab32158, Abcam). c-Fos and Bim primary antibodies were followed by a secondary horseradish peroxidase (HRP)-conjugated goat anti-rabbit IgG (H+L) antibody (A0208, Beyotime, Shanghai, China). β-actin was detected using beta Actin Antibody (HRP Conjugated) (AB2001, Abways Technology, Shanghai, China). Western blot data were analyzed using ImageJ software.

2.5. Quantitative PCR

The qPCR for validating the mRNA expression of *JUN*, *FOS*, and *Bim* in U251 (human malignant glioma cells) and LN229 cells (glioblastoma) was performed. In samples collected at 0 h after sham treatment, 0 h, 3 h, and 6 h after the electroporation, total RNA was extracted using a FastPure® Cell/Tissue Total RNA Isolation Kit (RC101, Vazyme Biotech Co., Ltd., Nanjing, Jiangsu, China) according to the manufacturer's instructions. First Strand cDNA was synthesized according to the instructions of the Thermo Scientific RevertAid First Strand cDNA Synthesis Kit (Thermo Scientific, USA). mRNA was quantified using the Bestar® SybrGreen qPCR mastermix kit (DBI® Bioscience, Ludwigshafen, German) according to the instruction manuals. The mRNA expression levels in the control groups were set at 1.00.

2.6. Statistics

In the present analysis, 12 samples (4 time points × 3 biological replicates) were collected for transcriptome sequencing. Raw reads were quality filtered using Trimmomatic to remove low-quality reads and obtain the clean reads (https://github.com/timflutre/trimmomatic (accessed on 20 May 2021)). Approximately 6.77~7.29 G of clean data for each sample were retained for subsequent analyses. The Q30 base distribution ranged from 94.57 to 94.83%. The average GC composition was 49.68 %. Reads were mapped to the human genome (GRCh38) using the HISAT2 (https://daehwankimlab.github.io/hisat2/ (accessed on 22 May 2021)). The alignment rate across the 12 samples to the reference genome ranged from 97.02 to 98.55%. On the basis of the alignments, the FPKM of each gene was obtained using Cufflinks [16], and read counts of each gene were acquired by HTSeqcount (https://htseq.readthedocs.io/en/master/ (accessed on 25 May 2021)). The q-value represents the corrected p-value using the Benjamini and Hochberg method [17]. The threshold for significantly differential expression was set at q-value < 0.05 and foldchange > 2 or foldchange < 0.5. Differential expression gene (DEGs) analysis was carried out using the DESeq (2012) R package (https://www.huber.embl.de/users/anders/DESeq/ (accessed on 25 May 2021)). Hierarchical cluster analysis of DEGs was performed to display the expression pattern of genes in different samples. GO enrichment and KEGG pathway enrichment analysis of DEGs were carried out, respectively, using R on the basis of the hypergeometric distribution. The possible RCD-related genes were screened by comparing DEGs to RCD gene lists from KEGG (Apoptosis: hsa04210, hsa04215; Autophagy: hsa04136, hsa04140; Necroptosis: hsa04217; Ferroptosis: hsa04216; Pyroptosis: not available). The potential immune-related genes were screened by comparing DEGs to gene lists from secondary classification terms (Immune disease and Immune system) based on KEGG. Human transcription factor names and their targets were obtained through hTFtarget (http://bioinfo.life.hust.edu.cn/hTFtarget#!/ (accessed on 5 October 2021)). The possible RCD-related transcription factors were screened by comparing the RCD-related genes from the present results with transcription factors from hTFtarget. Corresponding potential target genes were predicted by comparing transcription factor targets from hTFtarget with DEGs. An additional condition of searching the target genes and filtering pathways was set at gene expression levels (FPKM) ≥ 3 to avoid false-positive patterns. The STRING database (http://string-db.org (accessed on 9 October 2021)) was utilized to construct the PPI network of RCD-related DEGs, transcription factors, and corresponding target genes, with a combined score ≥ 0.4. The PPI data were demonstrated through Cytoscape (https://cytoscape.org/ (accessed on 9 October 2021)).

3. Results

3.1. Electric Field Intensity Threshold for Apoptosis/Necrosis of U251 Glioma Cell Line

With the field intensity incrementally increased, the percentages of the green/red fluorescent area in relation to the entire uniform electrical field-treated area sharply increased from basal values to near-constant values (Figure 1A). Fine S-shaped curves were obtained after the nonlinear sigmoid curve regression fitting. Apoptosis and necrosis were induced with a 50% response after electroporation treatment (Field intensity ED50 value) of 299.5 ± 1.7 V/cm and 542.6 ± 1.6 V/cm (mean ± Std. Error), respectively. The threshold field intensity (initial cell response) of electroporation-induced apoptosis and necrosis was approximately 270 V/cm and 520 V/cm, respectively. Representative fluorescence pictures of unaffected (240 V/cm) (blue fluorescence), apoptotic (520 V/cm) (green fluorescence), and necrotic cells (640 V/cm) (red fluorescence) are provided (Figure 1B). Incubation continued for a further 24 h after electroporation treatment. Corresponding light microscopy pictures of the three different states are demonstrated (Figure 1C). Field intensity below the threshold field intensity of apoptosis induction did not exhibit significant cytotoxicity on the U251 cell line. The unaffected cells showed no apparent changes in cell density and cellular morphology. After being exposed to the field intensity between the threshold field intensity of apoptosis induction and necrosis induction, the apoptotic cells exhibited drastically different behaviors. These behaviors included membrane blebs formation, separation from the surrounding cells, weak extracellular attachments to the plates, and disintegration of the cell into apoptotic bodies. Field intensity beyond the threshold field intensity of necrosis induction leads to the coagulative necrosis of cells. A similar phenomenon was also observed after the cells were exposed to the flat non-uniform electric fields generated by two parallel-needle electrodes (Figure 1D). Green fluorescence shows that the cells underwent apoptosis, and similar apoptotic morphologies appeared after a further 24 h incubation. The cells stained with red fluorescence appeared to have coagulative necrosis. Other areas show that the cells were intact, as observed 24 h later using light microscopy. The experimentally observed distribution shapes and boundaries of the three states are similar to numerical simulations (solid red and yellow lines). However, the simulation-predicted threshold field intensity of electroporation-induced apoptosis and necrosis seemed to be approximately 500 V/cm and 800 V/cm, respectively.

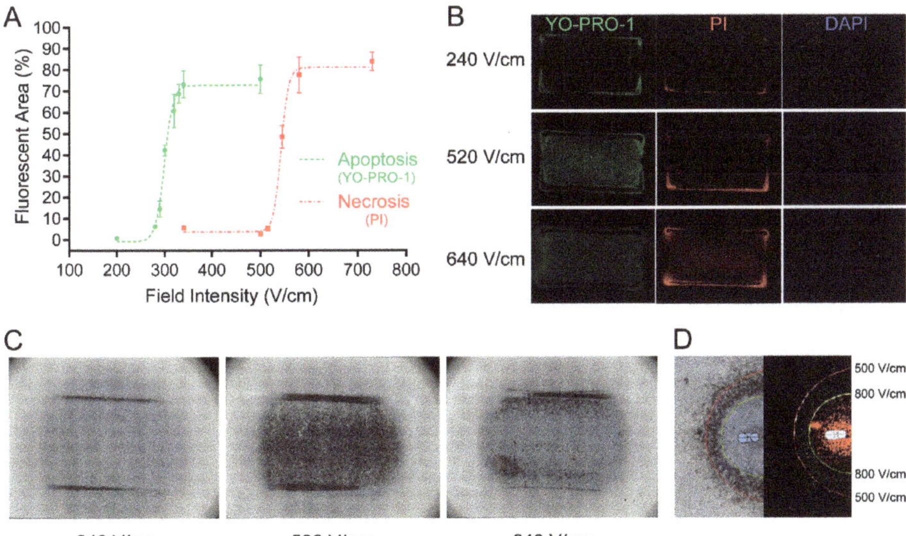

Figure 1. Electroporation-induced apoptosis and necrosis were determined by field intensity. (**A**) Field intensity of electroporation influenced the percentages of green (apoptosis) and red (necrosis) fluorescent areas to the entire treated area in a dose–effect-related manner. The threshold field intensity of electroporation-induced apoptosis and necrosis was approximately 270 V/cm and 520 V/cm, respectively. (**B**) Representative fluorescence images of unaffected (green and red fluorescence double negative), apoptotic (green fluorescence single positive), and necrotic cells (green and red fluorescence double positive) are presented. Nuclei of all cells were visualized using DAPI staining (blue fluorescence). (**C**) Corresponding representative light microscopy pictures of unaffected (240 V/cm), apoptotic (520 V/cm), and necrotic cells (640 V/cm) were acquired 24 h after electroporation. (**D**) Similar fluorescence after electroporation (right part) and corresponding morphological changes 24 h afterwards (Left part) were reproduced by the flat non-uniform electric fields generated by two parallel-needle electrodes. Solid red and yellow lines are numerical simulations of rough apoptotic and necrotic threshold field intensity (500 V/cm for the red line and 800 V/cm for the yellow line). Perspective blue and white cylinders represent one of the parallel-needle electrodes.

3.2. Comprehensive Regulatory Transcription Changes of IRE-Induced Apoptosis

Cells of apoptosis induction by IRE (520 V/cm) and sham intervention were sampled (0–6 h) for high-throughput transcriptome sequencing. There are 33 differential expression genes at 0 h after electroporation vs. 0 h after sham treatment (14 upregulated and 19 downregulated), 612 differential expression genes at 3 h (505 upregulated and 107 downregulated), and 1500 differential expression genes at 6 h (1223 upregulated and 277 downregulated). Volcano plots were employed to depict the differences in gene expression (Figure 2A). The results of the hierarchical cluster analysis are displayed in the form of a heat map (Figure 2B).

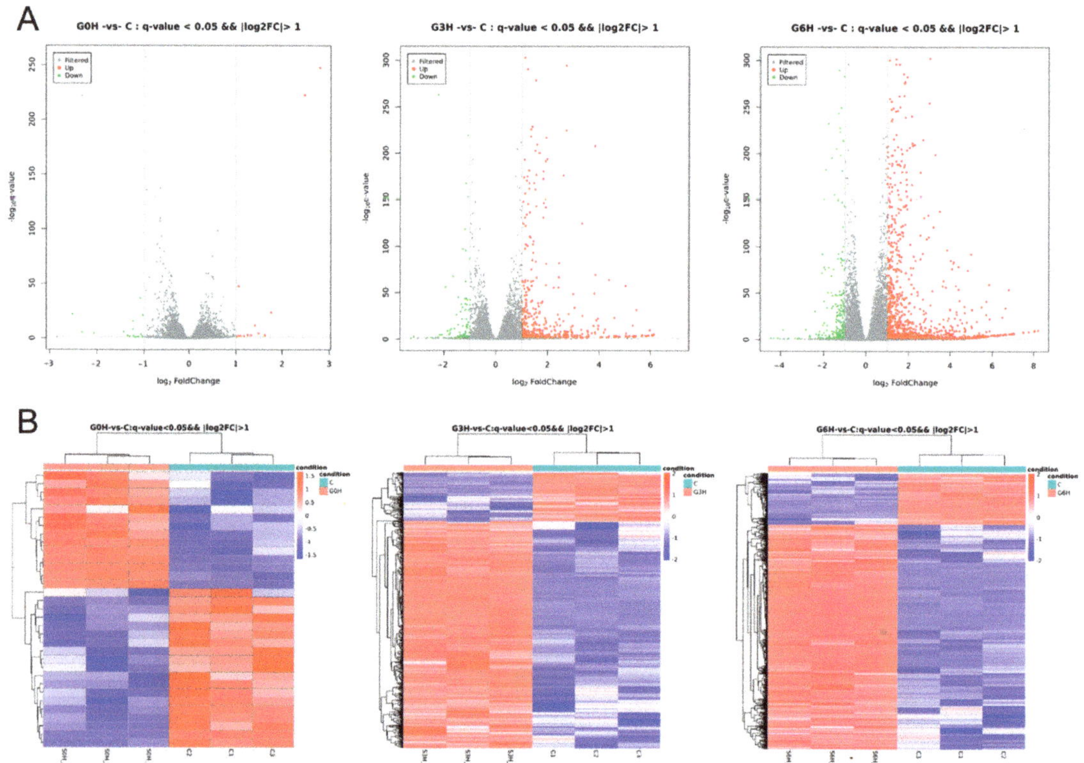

Figure 2. Time series transcriptomic changes of electroporation-induced apoptosis revealed gradually widespread gene expression differences. (**A**) Volcano plots for showing time series differentially expressed genes (*q*-value < 0.05 and foldchange > 2 or foldchange < 0.5). (**B**) Heat maps for displaying hierarchical cluster analysis of time series differentially expressed genes.

The up- and downregulated differential expression genes were annotated into 64 groups based on the GO level2 entries classifications (Figure 3A). GO annotations consist of three categories of differential expression genes: biological process, molecular function, and cellular component. The top 10 GO terms, sorted by the -log10P-value and containing at least three differential expression genes, in the three categories were screened out, respectively (Figure 3B). GO term "transcription factor *AP-1* complex" (black rectangle) of "cellular component" is top-ranked. Additionally, the GO term "inflammatory response" (black rectangle) of "biological process" is also in the ranking list. Bubble plots are used to portray the top 20 enriched KEGG pathways, sorted by the -log10P-value and containing at least three differential expression genes (Figure 4). Inflammatory-related signaling pathways

(IL-17 signaling pathway, MAPK signaling pathway, TNF signaling pathway, and NF-kappa B signaling pathway) and apoptosis-related signaling pathways (TNF signaling pathway and FoxO signaling pathway) are top-ranked.

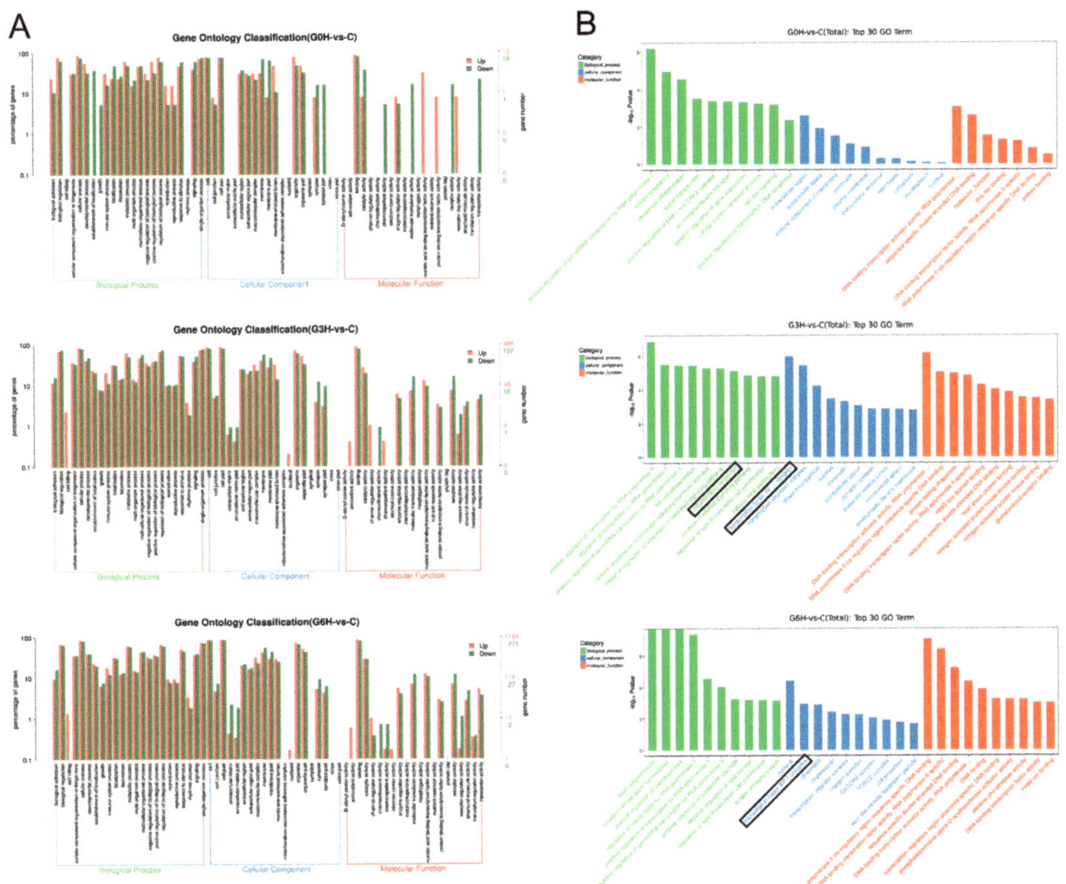

Figure 3. Data mining of the time series transcriptomic changes of electroporation-induced apoptosis. (**A**) Gene ontology (GO) term analysis was performed to annotate the up- and downregulated differential expression genes into 64 groups based on the GO level2 entries classifications. (**B**) GO terms were sorted by the -log10P-value, and the top 10 GO terms in each of the three GO level1 categories were displayed. The term "transcription factor *AP-1* complex" of "cellular component" and "inflammatory response" of "biological process" was depicted by a black rectangle.

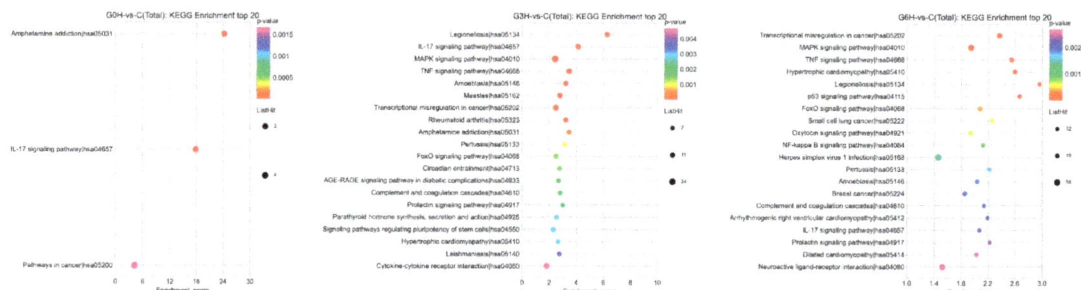

Figure 4. Enriched KEGG pathways were sorted by the -log10P-value, and the top 20 pathways were portrayed using bubble plots.

3.3. Transcription Factors FOS and JUN Played a Major Role in the Apoptosis after IRE, Accompanied by Mixed RCD Forms

In this experiment, a total of 31 possible RCD-related genes were screened out by comparing DEGs to RCD gene lists from KEGG (not counting ICD-related genes). Distributions of the 31 genes among different RCD forms, including apoptosis, autophagy, necroptosis, and ferroptosis, are shown in Table 1. However, there are no readily available pyroptosis pathway maps from KEGG. The role that pyroptosis plays in the RCD after IRE remains unclear. No differential expression genes in the present study were involved in the ferroptosis pathway.

Table 1. Distributions of the screening 31 possible RCD-related differential expression genes among RCD forms (not counting ICD-related genes).

Apoptosis	Apoptosis-Multiple Species	Autophagy-Other	Autophagy-Animal	Necroptosis	Pyroptosis	Ferroptosis
hsa04210	hsa04215	hsa04136	hsa04140	hsa04217	Not Available	hsa04216
BIRC3 BCL2L11 APAF1 PIK3R3 EIF2AK3 TNFRSF10A TNF	BIRC3 BCL2L11 APAF1		PIK3R3 EIF2AK3	BIRC3 TNFRSF10A TNF	N/A	/
		ATG9B	ATG9B DDIT4 DEPTOR			
GADD45B MCL1 JUN DDIT3 PIDD1 GADD45A NFKBIA FOS NTRK1 GADD45G ATF4						
				H2AC6 PLA2G4C H2AC20 PYGM JAK3 PLA2G4F IL1A USP21 TNFAIP3 ZBP1		

KEGG pathway names and entries were listed at the head of the table. "N/A" represented that there were no currently available pyroptosis pathway maps from KEGG. "/" represented that there was no distribution of RCD-related DEGs in "Ferroptosis" in this experiment.

Among the 31 possible RCD-related genes, the screening identified two transcription factors: *FOS* and *JUN*. PPI data of the union of the 31 RCD-related genes, two transcription factors, and corresponding transcription factors target genes, were demonstrated (Figure 5). The foldchange of *JUN* was 1.43 at 0 h after electroporation (mean expression level 115) vs.

0 h after sham treatment (mean expression level 78) and did not meet the predetermined threshold for significantly differential expression (foldchange > 2 or foldchange < 0.5). This question aside, *FOS*, in conjunction with *JUN*, played a central role in the RCD, including apoptosis at all three time points after IRE.

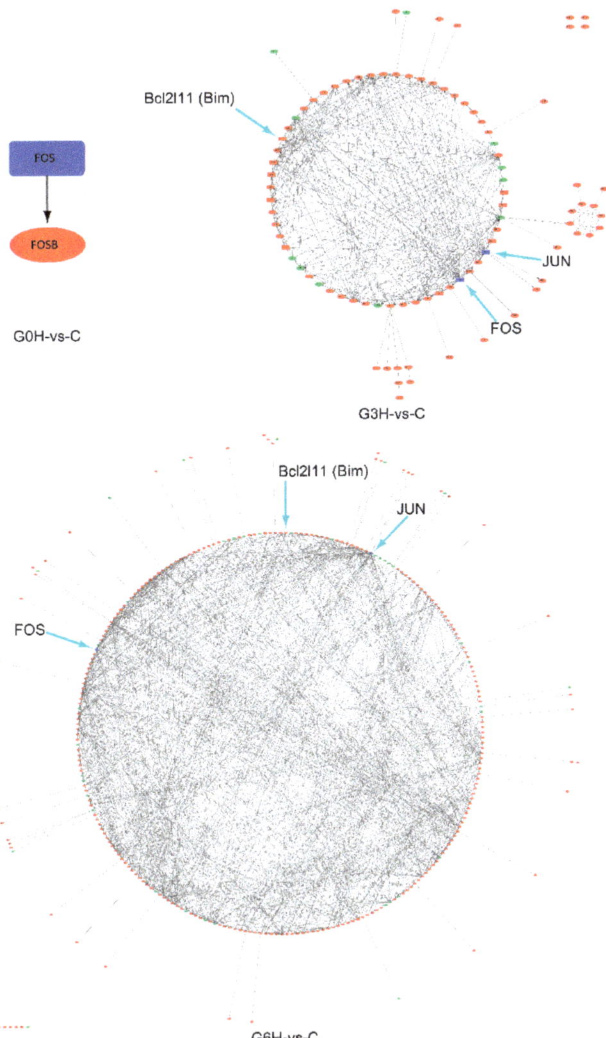

Figure 5. Time series differential expression genes interactions network of the 31 RCD-related genes (not counting ICD-related genes), 2 transcription factors, and corresponding target genes of the transcription factors. Blue nodes represent upregulated transcription factors; red nodes are upregulated genes; green nodes are downregulated genes; round rectangle nodes are RCD-related genes; ellipse nodes are non-RCD-related genes; edges are genes interactions; edge directions are the regulatory interactions from transcription factors to their target genes. Cyan arrowheads indicate amplified gene names on the gene interaction networks.

A total of 122 potential immune-related genes were screened by comparing DEGs and gene lists from secondary classification terms (immune disease and immune system) based on KEGG. The temporal profile of the immune-related differentially expressed genes of all three time points after IRE is shown in Table 2.

Table 2. Time course transcriptomic changes of electroporation-induced RCD to involve immune-related differentially expressed genes.

G0H-vs-C	G3H-vs-C	G6H-vs-C
FOS FOSB MMP1	FOS FOSB MMP1 FERMT3 NFKBIA ITGAM MAPK13 H2BU1 ATP6V1B1 GRIN2B C5AR1 HSPA6 EGR3 SERPINC1 TNFAIP3 H2BC18 CCL2 IL17B TNFRSF10A CXCL3 C3AR1 DLL1 MYLK3 IL6 PRKCG MAP3K8 PTPN6 IL23A PECAM1 CXCL2 EGR2 BCL3 IL1A ITGAX H3C4 IL17C C9 JUN CD1D CD79A BUB1B-PAK6 IL1R2 GNB3 RASGRP2 H2BC17 JAK3 HSPA1B PLA2G4F RASSF5 CD14 HSPA1A	FOS FOSB FERMT3 NFKBIA ITGAM MAPK13 H2BU1 ATP6V1B1 GRIN2B C5AR1 HSPA6 EGR3 SERPINC1 TNFAIP3 H2BC18 CCL2 IL17B TNFRSF10A CXCL3 C3AR1 DLL1 MYLK3 IL6 PRKCG MAP3K8 PTPN6 IL23A PECAM1 CXCL2 EGR2 BCL3 IL1A ITGAX H3C4 IL17C C9 JUN CD1D CD79A BUB1B-PAK6 IL1R2 GNB3 RASGRP2 H2BC17 JAK3 HSPA1B PLA2G4F RASSF5 CD14 HSPA1A
CCL26	HSPA2 FGG COL3A1 F3 CD74 APBB1IP	KIT TNFRSF13C H2BC6 H2BC5 H2BC4 CD34 C2 TBX21 DLL3 H3C13 SCIN CXCR4 MASP2 GRK7 GBP1 PLA2G4C FOSL1 KSR1 H2AC20 MYLK4 F2 FOXO3 H2BC11 POLR1C TNFRSF11A RUNX1 CD1A PTGS2 BDKRB2 CALML4 TNF FLT1 LCK TLR9 H2AC6 RARA DLL4 GUCY1A2 NFATC2 CLDN1 TBKBP1 TLR5 IL5RA F2RL2 POLR2H RAG1 IRF1 F2R CLDN19 NFKBIE IRF3 SERPINF2 PIK3R3 BIRC3 JAM2 THBD ITGB7 H3C3 ITGA2B H2BC8 NAMPT ZBP1 PIP5K1A BTK

Head of the table shows the sampling time points. Red font denotes transcription factors.

Finally, we confirmed the sequencing results by qPCR and Western blot (Figure 6). The mRNA levels of *FOS* and *JUN*, major constituents of the *AP-1* transcription factor, and *Bim*, a known executioner of RCD, were all gradually upregulated after the treatment (Figure 6C–H). Expression of the protein c-Fos was also consistently upregulated (Figure 6A,B). These all echoed the sequencing results. However, the expression of the protein Bim was complicated (Figure 6A,B). Bim showed high basal expression and did not reduce instantly after the treatment. The protein amount of Bim then dropped to the lowest expression at 3 h and gradually upregulated 6 h later. This will be discussed in the 'Discussion' section. Overall, the qPCR and Western blot findings supported the sequencing results.

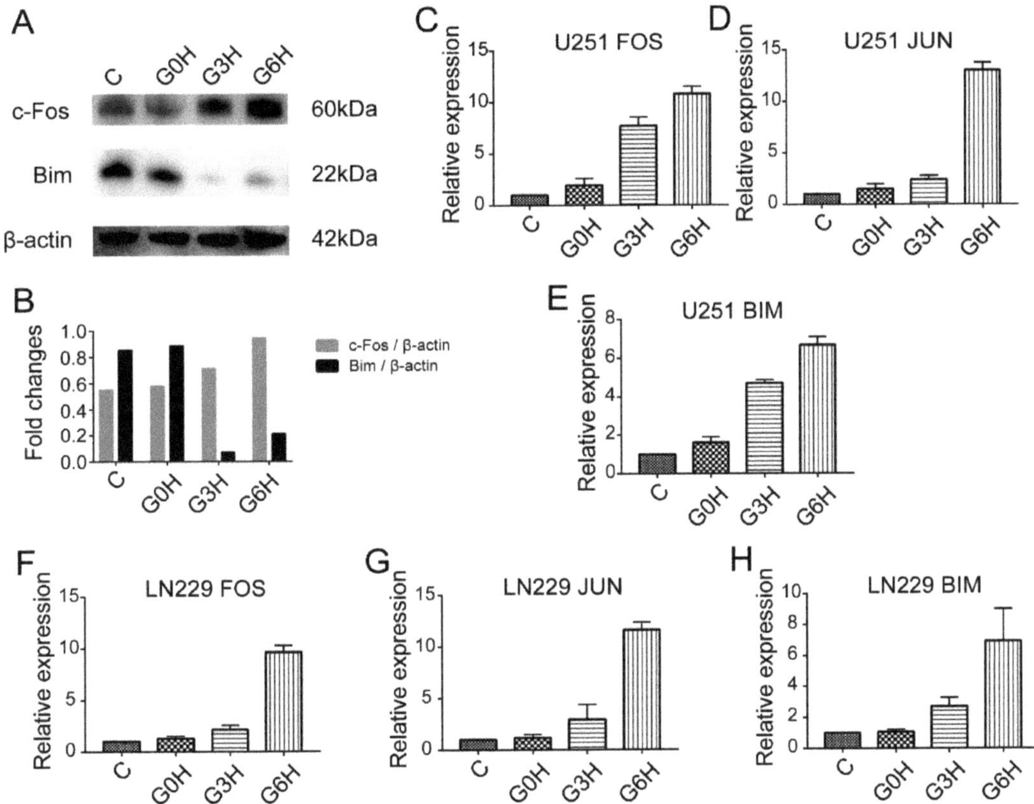

Figure 6. Upregulation of transcription factor *AP-1* and upregulation of *Bim* (Bcl2l11) expression after electroporation. Western blot results (**A**) and analysis (**B**) of *c-Fos* and *Bim* in samples collected at 0 h after sham treatment, 0 h, and 6 h after electroporation in U251 cells. *c-Fos* was consistently upregulated. *Bim* showed high basal expression and did not significantly reduce instantly after the treatment and then dropped to the lowest expression 3 h later. Expression of *Bim* was gradually upregulated 6 h later. qPCR results (**C–H**) of *FOS* (**C,F**), *JUN* (**D,G**), and *Bim* (**E,H**) in two different cells, U251 (**C–E**) and LN229 cells (**F–H**), all showed gradual upregulation after electroporation, which was consistent with the sequencing results.

4. Discussion

The primary purpose of the present study was to comprehensively explore the mechanism of apoptosis after IRE based on high-throughput transcriptome sequencing. To confirm apoptosis, YO-PRO-1/PI co-staining and cellular morphological changes after re-culture for another 24 h were used (Figure 1B–D). YO-PRO-1 staining confirmed apoptosis in near real-time [18]. Cellular changes of typical apoptotic morphologies [19] 24 h later, including membrane blebs formation, separation from the surrounding cells, weak extracellular attachments to the plates, and disintegration of the cell into apoptotic bodies, further validated apoptosis.

However, the fact that COMSOL simulation-predicted threshold field intensity of electroporation-induced apoptosis and necrosis was approximately 500 V/cm and 800 V/cm, respectively, needs to be discussed in depth. We noticed that the cathode electrode of the two parallel-needle electrodes generated a large number of gas bubbles (Figure S1), while a similar phenomenon was not apparent on the two parallel-plate electrodes. The gas bubbles formation may arise from the excessive field strength generated around the needle electrodes. Most of the cathode electrode surface of the needle electrode was covered by the bubbles, causing local current density to increase, leading to local field intensity increasing, resulting in the experimental field strengths being smaller than the predicted values in other regions. The difference between the simulation-predicted apoptosis and necrosis threshold field intensity of electroporation (approximately 300 V/cm) is similar to the difference between the field intensity ED50 value for apoptosis and necrosis induction (243.1 V/cm). The overpredicted field strengths are likely to arise from gas bubble formation on the needle electrode. Field intensity ED50 values for apoptosis and necrosis induction obtained by utilizing the two parallel-plate electrodes are more robust. Like our results, electrode design for IRE should adequately consider the influence of bubble formation [20].

Through the optimized field intensity gradient settings of flat uniform electric fields, fine S-shaped curves were obtained (Figure 1A), which is similar to "dose-effect relation-ships" in pharmacological investigations. Apoptosis and necrosis sharply increased in a field intensity-dependent manner if the corresponding thresholds were exceeded. Origi-nally, ED 50 was defined as the median effective dose, the dose of a drug that is effective in 50% of individuals [21]. Field intensity resembles the dose of the drug, and individuals refer to single cells here. For the first time, we draw on the pharmacological concepts of "S-shaped curves", "dose-effect relationships", and "ED50" to study the biological effects of electroporation.

The number of differentially expressed genes from samples after apoptosis induction by IRE increased progressively with time from 0 h to 6 h in a regulatory way. Profiles of transcrip-tome sequencing were unbiasedly demonstrated (Figure 2A,B, Figure 3A,B and Figure 4). The "transcription factor *AP-1* complex", immune-related go terms, and signaling path-ways were more prominent. Firstly, activator protein-1 (*AP-1*) is a classical regulator of cellular apoptosis in specific cell types [22]. *AP-1* consists of heterodimers and homodimers of *JUN*, *FOS*, or activating transcription factor bZIP proteins. However, to our knowledge, this is the first report of such a transcription factor in the gene regulatory mechanism of apoptosis after IRE. Secondly, what deserves attention is that apoptosis is usually immunologically silent [23], contrary to our transcriptome sequencing results, that 122 potential immune-related genes were involved. In-depth data mining for other forms of RCD [14] will be performed later.

Possible differentially expressed RCD-related genes were screened out (Table 1). Al-though iron-based stainless-steel electrodes might generate in situ Fe^{2+} during electropo-ration, which plays an important role in ferroptosis [24], no differential expression genes were detected to be involved in the ferroptosis pathway in the present study. Involve-ment of ferroptosis in RCD within 6 h after IRE can therefore be ruled out. The possible RCD-related genes are distributed in different RCD forms, including apoptosis, autophagy, and necroptosis. Necroptosis is a highly immunogenic type of RCD [25], and autophagy has been reported to control further inflammatory responses [26]. Apoptosis is linked to DNA damage via several pathways that were not yet fully studied in the case of nsEP. One possibility is via PLK-1 protein and centrosome-mediated apoptosis, PUMA and NOXA were not activated. Exposure of cells to nsEP causes ER stress that could be related to ROS formation, permeabilization of the ER, or Ca^{2+} influx and can further trigger mitochondria-mediated intrinsic apoptosis via PERK and IRE1. ROS formation that occurs after nsEP could trigger both intrinsic and extrinsic apoptosis via several cellular targets, including those in the mitochondria, DNA, ER, and plasma membrane [13]. Among the 122 poten-tial immune-related genes, 28 genes act as pro-inflammatory factors, and 4 genes act as anti-inflammatory factors [27]. The 28 pro-inflammatory genes we detected are 14 differ-entially expressed pro-inflammatory cytokines, cytokine receptors, and cytokine-induced proteins: *IL17B, IL17C, IL1A, IL23A, IL6, IL5RA, TNF, TNFRSF11A, TNFRSF13C, CCL2, CCL26, CXCL2, CXCL3,* and TNF alpha-induced protein 3 (*TNFAIP3*); pro-inflammatory complement receptors *C3AR1* and *C5AR1*; heat shock proteins *HSPA1A* and *HSPA1B*; Toll-like receptors *TLR5* and *TLR9*; leukocyte differentiation antigen: *CD14, CD1A, CD1D, CD74,* and *CD79A*; kinases *JAK3* and *MAPK13*; nuclear factor *NFATC2*. On the opposite side, four anti-inflammatory factors, including cytokine receptors (*TNFRSF10A* and *IL1R2*) and NFKB inhibitors (*NFKBIE* and *NFKBIA*), were detected. Although mainly produced by immune cells, inflammation-related substances produced by tumor cells are also common [28]. More promisingly, these inflammation-related substances after IRE endow the U251 glioma cells with the potential of generating a systemic antitumor immune response. The immunogenicity merits further thorough inquiry. As amply discussed above, the apoptosis induction of IRE caused multiple RCD forms (involvement of apoptosis, autophagy, necroptosis, and ICD transcriptome mechanisms) simultaneously.

Except for the fact that the foldchange of *JUN* at 0 h after the electroporation (mean expression level at 0 h after sham treatment as benchmark) did not meet the predetermined threshold for significantly differential expression, upregulation of *FOS* and *JUN*, major constituents of the *AP*-ranscription factor [22], maintained throughout all the three time points (0 h, 3 h, and 6 h). *AP-1* transcription factor was also the only transcription factor among the 31 possible RCD-related genes. GO term "transcription factor *AP-1* complex" of "cellular component" was top-ranked after GO enrichment analysis of all DEGs (Figure 3B). Electrical stimulation was reported to induce the upregulation of *AP-1* [29]. Additionally, increased *AP-1* activity can cause apoptosis in specific cell types, including tumor cells and neuronal cells [22]. Similarly, transcription factor *AP-1* should play a key role during apoptosis after the irreversible electroporation in this experiment. However, *FOS* expression duration and intensity in our study were considerably higher than the moderate one in the published data [29]. Su et al. reported a transient *FOS* expression elevation 1 h after electro-stimulation (approximate expression level = 10, fold change = 2), returning to baseline after 4 h, causing no significant cell death and excitotoxicity. Our results show an immediately early upregulation after the irreversible electroporation, a drastic upregulation in less than 3 h, and maintenance of high expression level for at least 6 h. It is widely believed that transcription factor *AP-1* mediates apoptosis by mainly regulating the gene expression of *P53* [30,31], *Fas* [32], *Fas-L* [32], *Bim* [33], and *HRK* [34]. Combining our results, *Bim*, also named Bcl2l11, was the only DEG with gene expression levels (FPKM) ≥ 3, whose expression level gradually increased and significantly upregulated at 3 h and 6 h. Putcha et al. [35] reported that the protein amount of *Bim* started to gradually upregulate 6 h after a cellular disturbance. The trend toward the upregulation of *Bim* from 3 h to 6 h (Figure 6A,B) after electroporation was consistent with the findings of Putcha et al. Importantly and different from them, the U251 cells showed high basal expression of *Bim* at 0 h after sham treatment (C group) and 0 h after electroporation (G0H group) (Figure 6A,B). We conjectured that the lowest expression of *Bim* at 3 h was due to protein leakage of basal *Bim* through cellular membrane holes formed after the electroporation. The upregulation of *Bim* around 6 h reflected de novo synthesis of protein *Bim*. However, more additional follow-up experiments need to be performed in the future.

In summary, our time-course transcriptome sequencing data suggest that IRE possibly mediate apoptosis by long-term high-intensity upregulation of transcription factor *AP-1* and upregulation of *Bim* (Bcl2l11) expression. Based on this prediction, further studies will need to be carried out on the pathways of IRE-induced apoptosis.

Supplementary Materials: The following supporting information can be downloaded at: https://www.mdpi.com/article/10.3390/brainsci12111465/s1, Figure S1. Light microscopy picture of gas bubbles formation around the cathodal needle electrode. Red dotted ellipse indicates the electrode (Diameter: 0.4 mm). Red arrows indicate gas bubbles.

Author Contributions: Conceptualization, B.Z. and Z.Q.; Data curation, S.Y. and L.C.; Formal analysis, L.C.; Funding acquisition, B.Z. and Z.Q.; Investigation, S.Y.; Methodology, S.Y., K.S., T.S., Z.F., L.D., J.L., L.J. and G.Z.; Project administration, S.Y.; Supervision, B.Z. and Z.Q.; Validation, S.Y.; Writing—original draft, S.Y.; Writing—review and editing, L.C. All authors have read and agreed to the published version of the manuscript.

Funding: This work was partially supported by grant from the National Natural Science Foundation of China to the corresponding author (BZ) (Grant No. 81801795). This work was partially supported by the CSCO treatment and Research Fund Project (Y-zai2021/ms-0270 and Y-zai2021/qn-0204).

Institutional Review Board Statement: Not applicable.

Data Availability Statement: Data can be available upon request by contacting the corresponding author via email.

Conflicts of Interest: The authors declare that no conflict of interest exist.

References

1. Ostrom, Q.T.; Gittleman, H.; Liao, P.; Vecchione-Koval, T.; Wolinsky, Y.; Kruchko, C.; Barnholtz-Sloan, J.S. CBTRUS Statistical Report: Primary brain and other central nervous system tumors diagnosed in the United States in 2010–2014. *Neuro-Oncology* **2017**, *19*, V1–V88. [CrossRef]
2. Suchorska, B.; Weller, M.; Tabatabai, G.; Senft, C.; Hau, P.; Sabel, M.C.; Herrlinger, U.; Ketter, R.; Schlegel, U.; Marosi, C.; et al. Complete resection of contrast-enhancing tumor volume is associated with improved survival in recurrent glioblastoma-results from the DIRECTOR trial. *Neuro-Oncology* **2016**, *18*, 549–556. [CrossRef]
3. Pasqualetti, F.; Montemurro, N.; Desideri, I.; Loi, M.; Giannini, N.; Gadducci, G.; Malfatti, G.; Cantarella, M.; Gonnelli, A.; Montrone, S.; et al. Impact of recurrence pattern in patients undergoing a second surgery for recurrent glioblastoma. *Acta Neurol. Belg.* **2022**, *122*, 441–446. [CrossRef]
4. Stupp, R.; Mason, W.P.; van den Bent, M.J.; Weller, M.; Fisher, B.; Taphoorn, M.J.B.; Belanger, K.; Brandes, A.A.; Marosi, C.; Bogdahn, U.; et al. Radiotherapy plus concomitant and adjuvant temozolomide for glioblastoma. *N. Engl. J. Med.* **2005**, *352*, 987–996. [CrossRef]
5. Niessen, C.; Thumann, S.; Beyer, L.; Pregler, B.; Kramer, J.; Lang, S.; Teufel, A.; Jung, E.M.; Stroszczynski, C.; Wiggermann, P. Percutaneous Irreversible Electroporation: Long-term survival analysis of 71 patients with inoperable malignant hepatic tumors. *Sci. Rep.* **2017**, *7*, 43687. [CrossRef]
6. Paiella, S.; Butturini, G.; Frigerio, I.; Salvia, R.; Armatura, G.; Bacchion, M.; Fontana, M.; D'Onofrio, M.; Martone, E.; Bassi, C. Safety and Feasibility of Irreversible Electroporation (IRE) in Patients with Locally Advanced Pancreatic Cancer: Results of a Prospective Study. *Dig. Surg.* **2015**, *32*, 90–97. [CrossRef]
7. Valerio, M.; Dickinson, L.; Ali, A.; Ramachandran, N.; Donaldson, I.; Freeman, A.; Ahmed, H.U.; Emberton, M. A prospective development study investigating focal irreversible electroporation in men with localised prostate cancer: Nanoknife Electroporation Ablation Trial (NEAT). *Contemp. Clin. Trials* **2014**, *39*, 57–65. [CrossRef]
8. Wendler, J.J.; Porsch, M.; Nitschke, S.; Kollermann, J.; Siedentopf, S.; Pech, M.; Fischbach, F.; Ricke, J.; Schostak, M.; Liehr, U.B. A prospective Phase 2a pilot study investigating focal percutaneous irreversible electroporation (IRE) ablation by NanoKnife in patients with localised renal cell carcinoma (RCC) with delayed interval tumour resection (IRENE trial). *Contemp. Clin. Trials* **2015**, *43*, 10–19. [CrossRef]
9. Charpentier, K.P.; Wolf, F.; Noble, L.; Winn, B.; Resnick, M.; Dupuy, D.E. Irreversible electroporation of the liver and liver hilum in swine. *HPB* **2011**, *13*, 168–173. [CrossRef]
10. Fang, Z.; Chen, L.C.; Moser, M.A.J.; Zhang, W.J.; Qin, Z.Y.; Zhang, B. Electroporation-Based Therapy for Brain Tumors: A Review. *J. Biomech. Eng.-Trans. Asme* **2021**, *143*, 100802. [CrossRef]
11. Rossmeisl, J.H.; Garcia, P.A.; Pancotto, T.E.; Robertson, J.L.; Henao-Guerrero, N.; Neal, R.E.; Ellis, T.L.; Davalos, R.V. Safety and feasibility of the Nano Knife system for irreversible electroporation ablative treatment of canine spontaneous intracranial gliomas. *J. Neurosurg.* **2015**, *123*, 1008–1025. [CrossRef]
12. Shu, T.; Ding, L.; Fang, Z.; Yu, S.; Chen, L.; Moser, M.; Zhang, W.J.C.; Qin, Z.; Zhang, B. Lethal Electric Field Thresholds for Cerebral Cells with IRE and H-FIRE Protocols: An In Vitro 3D Cell Model Study. *J. Biomech. Eng.* **2022**, *144*, 101010. [CrossRef]
13. Napotnik, T.B.; Polajzer, T.; Miklavcic, D. Cell death due to electroporation—A review. *Bioelectrochemistry* **2021**, *141*, 107871. [CrossRef]
14. Galluzzi, L.; Vitale, I.; Aaronson, S.A.; Abrams, J.M.; Adam, D.; Agostinis, P.; Alnemri, E.S.; Altucci, L.; Amelio, I.; Andrews, D.W.; et al. Molecular mechanisms of cell death: Recommendations of the Nomenclature Committee on Cell Death 2018. *Cell Death Differ.* **2018**, *25*, 486–541. [CrossRef]
15. Idziorek, T.; Estaquier, J.; Debels, F.; Ameisen, J.C. Yopro-1 Permits Cytofluorometric Analysis Of Programmed Cell-Death (Apoptosis) Without Interfering With Cell Viability. *J. Immunol. Methods* **1995**, *185*, 249–258. [CrossRef]
16. Trapnell, C.; Williams, B.A.; Pertea, G.; Mortazavi, A.; Kwan, G.; van Baren, M.J.; Salzberg, S.L.; Wold, B.J.; Pachter, L. Transcript assembly and quantification by RNA-Seq reveals unannotated transcripts and isoform switching during cell differentiation. *Nat. Biotechnol.* **2010**, *28*, 511–515. [CrossRef]
17. Benjamini, Y.; Hochberg, Y. Controlling The False Discovery Rate—A Practical And Powerful Approach To Multiple Testing. *J. R. Stat. Soc. Ser. B Methodol.* **1995**, *57*, 289–300. [CrossRef]
18. Chen, C.L.; Kuo, L.R.; Chang, C.L.; Hwu, Y.K.; Huang, C.K.; Lee, S.Y.; Chen, K.; Lin, S.J.; Huang, J.D.; Chen, Y.Y. In situ real-time investigation of cancer cell photothermolysis mediated by excited gold nanorod surface plasmons. *Biomaterials* **2010**, *31*, 4104–4112. [CrossRef]
19. Hacker, G. The morphology of apoptosis. *Cell Tissue Res.* **2000**, *301*, 5–17. [CrossRef]
20. Wojtaszczyk, A.; Caluori, G.; Pešl, M.; Melajova, K.; Stárek, Z. Irreversible electroporation ablation for atrial fibrillation. *J. Cardiovasc. Electrophysiol.* **2018**, *29*, 643–651. [CrossRef]

21. Litchfield, J.T., Jr.; Wilcoxon, F. A simplified method of evaluating dose-effect experiments. *J. Pharmacol. Exp. Ther.* **1949**, *96*, 99–113.
22. Eferl, R.; Wagner, E.F. AP-1: A double-edged sword in tumorigenesis. *Nat. Rev. Cancer* **2003**, *3*, 859–868. [CrossRef]
23. Voll, R.E.; Herrmann, M.; Roth, E.A.; Stach, C.; Kalden, J.R.; Girkontaite, I. Immunosuppressive effects of apoptotic cells. *Nature* **1997**, *390*, 350–351. [CrossRef]
24. Dixon, S.J.; Lemberg, K.M.; Lamprecht, M.R.; Skouta, R.; Zaitsev, E.M.; Gleason, C.E.; Patel, D.N.; Bauer, A.J.; Cantley, A.M.; Yang, W.S.; et al. Ferroptosis: An iron-dependent form of nonapoptotic cell death. *Cell* **2012**, *149*, 1060–1072. [CrossRef]
25. Snyder, A.G.; Hubbard, N.W.; Messmer, M.N.; Kofman, S.B.; Hagan, C.E.; Orozco, S.L.; Chiang, K.; Daniels, B.P.; Baker, D.; Oberst, A. Intratumoral activation of the necroptotic pathway components RIPK1 and RIPK3 potentiates antitumor immunity. *Sci. Immunol.* **2019**, *4*, eaaw2004. [CrossRef]
26. Zhong, Z.; Sanchez-Lopez, E.; Karin, M. Autophagy, Inflammation, and Immunity: A Troika Governing Cancer and Its Treatment. *Cell* **2016**, *166*, 288–298. [CrossRef]
27. Dinarello, C.A. Anti-inflammatory Agents: Present and Future. *Cell* **2010**, *140*, 935–950. [CrossRef]
28. Poggi, A.; Zocchi, M.R. Mechanisms of tumor escape: Role of tumor microenvironment in inducing apoptosis of cytolytic effector cells. *Arch. Immunol. Ther. Exp.* **2006**, *54*, 323–333. [CrossRef]
29. Su, Y.; Shin, J.; Zhong, C.; Wang, S.; Roychowdhury, P.; Lim, J.; Kim, D.; Ming, G.L.; Song, H. Neuronal activity modifies the chromatin accessibility landscape in the adult brain. *Nat. Neurosci.* **2017**, *20*, 476–483. [CrossRef]
30. Kolev, V.; Mandinova, A.; Guinea-Viniegra, J.; Hu, B.; Lefort, K.; Lambertini, C.; Neel, V.; Dummer, R.; Wagner, E.F.; Dotto, G.P. EGFR signalling as a negative regulator of Notch1 gene transcription and function in proliferating keratinocytes and cancer. *Nat. Cell Biol.* **2008**, *10*, 902–911. [CrossRef]
31. Hu, W.; Chan, C.S.; Wu, R.; Zhang, C.; Sun, Y.; Song, J.S.; Tang, L.H.; Levine, A.J.; Feng, Z. Negative regulation of tumor suppressor p53 by microRNA miR-504. *Mol. Cell.* **2010**, *38*, 689–699. [CrossRef]
32. Kasibhatla, S.; Brunner, T.; Genestier, L.; Echeverri, F.; Mahboubi, A.; Green, D.R. DNA damaging agents induce expression of Fas ligand and subsequent apoptosis in T lymphocytes via the activation of NF-kappa B and AP-1. *Mol. Cell* **1998**, *1*, 543–551. [CrossRef]
33. Whitfield, J.; Neame, S.J.; Paquet, L.; Bernard, O.; Ham, J. Dominant-negative c-Jun promotes neuronal survival by reducing BIM expression and inhibiting mitochondrial cytochrome c release. *Neuron* **2001**, *29*, 629–643. [CrossRef]
34. Chen, S.W.; Lee, J.M.; Zeng, C.B.; Chen, H.; Hsu, C.Y.; Xu, J. Amyloid beta peptide increases DP5 expression via activation of neutral sphingomyelinase and JNK in oligodendrocytes. *J. Neurochem.* **2006**, *97*, 631–640. [CrossRef]
35. Putcha, G.V.; Le, S.; Frank, S.; Besirli, C.G.; Clark, K.; Chu, B.; Alix, S.; Youle, R.J.; LaMarche, A.; Maroney, A.C.; et al. JNK-mediated BIM phosphorylation potentiates BAX-dependent apoptosis. *Neuron* **2003**, *38*, 899–914. [CrossRef]

Article

Multiparametric Longitudinal Profiling of RCAS-tva-Induced PDGFB-Driven Experimental Glioma

Hannes Becker [1,2,*,†], Salvador Castaneda-Vega [3,4,†], Kristin Patzwaldt [3], Justyna M. Przystal [1,5], Bianca Walter [1], Filippo C. Michelotti [3], Denis Canjuga [1], Marcos Tatagiba [1,2], Bernd Pichler [4,5,6], Susanne C. Beck [1], Eric C. Holland [7], Christian la Fougère [4,5,6] and Ghazaleh Tabatabai [1,5,6]

1. Department of Neurology & Interdisciplinary Neuro-Oncology, Hertie Institute for Clinical Brain Research, Center for Neuro-Oncology, Comprehensive Cancer Center, University Hospital Tübingen, Eberhard Karls University Tubingen, 72072 Tubingen, Germany
2. Department of Neurosurgery, University Hospital Tubingen, Eberhard Karls University Tubingen, 72072 Tubingen, Germany
3. Werner Siemens Imaging Center, Department of Preclinical Imaging and Radiopharmacy, Eberhard Karls University Tuebingen, 72072 Tubingen, Germany
4. Department of Nuclear Medicine and Clinical Molecular Imaging, Eberhard Karls University Tuebingen, 72072 Tubingen, Germany
5. German Translational Cancer Consortium (DKTK), DKFZ Partner Site, 72072 Tubingen, Germany
6. Cluster of Excellence iFIT (EXC 2180) "Image Guided and Functionally Instructed Tumor Therapies", Eberhard Karls University, 72072 Tubingen, Germany
7. Human Biology Division, Fred Hutchinson Cancer Research Center, Seattle, Washington, DC 98109, USA
* Correspondence: hannes.becker@med.uni-tuebingen.de; Tel.: +49(0)7071-2968676; Fax: 49(0)7071-295957
† These authors contributed equally to this work.

Citation: Becker, H.; Castaneda-Vega, S.; Patzwaldt, K.; Przystal, J.M.; Walter, B.; Michelotti, F.C.; Canjuga, D.; Tatagiba, M.; Pichler, B.; Beck, S.C.; et al. Multiparametric Longitudinal Profiling of RCAS-tva-Induced PDGFB-Driven Experimental Glioma. *Brain Sci.* **2022**, *12*, 1426. https://doi.org/10.3390/brainsci12111426

Academic Editor: Hailiang Tang

Received: 2 September 2022
Accepted: 19 October 2022
Published: 24 October 2022

Publisher's Note: MDPI stays neutral with regard to jurisdictional claims in published maps and institutional affiliations.

Abstract: Glioblastomas are incurable primary brain tumors harboring a heterogeneous landscape of genetic and metabolic alterations. Longitudinal imaging by MRI and [^{18}F]FET-PET measurements enable us to visualize the features of evolving tumors in a dynamic manner. Yet, close-meshed longitudinal imaging time points for characterizing temporal and spatial metabolic alterations during tumor evolution in patients is not feasible because patients usually present with already established tumors. The replication-competent avian sarcoma-leukosis virus (RCAS)/tumor virus receptor-A (tva) system is a powerful preclinical glioma model offering a high grade of spatial and temporal control of somatic gene delivery in vivo. Consequently, here, we aimed at using MRI and [^{18}F]FET-PET to identify typical neuroimaging characteristics of the platelet-derived growth factor B (PDGFB)-driven glioma model using the RCAS-tva system. Our study showed that this preclinical glioma model displays MRI and [^{18}F]FET-PET features that highly resemble the corresponding established human disease, emphasizing the high translational relevance of this experimental model. Furthermore, our investigations unravel exponential growth dynamics and a model-specific tumor microenvironment, as assessed by histology and immunochemistry. Taken together, our study provides further insights into this preclinical model and advocates for the imaging-stratified design of preclinical therapeutic interventions.

Keywords: glioblastoma; RCAS-tva; rodent glioma models; metabolism; multiparametric PET/MR imaging; model characterization; PDGFB

1. Introduction

A glioblastoma is one of the most aggressive primary tumors in the central nervous system. Despite multimodal therapy approaches, glioblastomas still have a devastating prognosis, with a median overall survival in the range of 1.5 years [1–5]. Standard of care therapy outside of clinical trials consists of maximal safe resection followed by a combination of radiation therapy, alkylating chemotherapy, and tumor-treating fields [1,3,6–8]. Still, tumor progression is inevitable in most patients. Re-resection can be considered if the whole contrast-enhancing tumor volume can be removed without new neurological symptoms [7,9,10]. Furthermore, alkylating chemotherapy or other systemic therapies are considered in interdisciplinary tumor boards [7,9–11].

Comprehensive genetic sequencing efforts over the last decades have revealed a complex network of mutations in human cancer entities [12]. Multi-omic approaches and extensive bioinformatic clustering analysis have enabled the definition of progressively sharper subsets of different cancer entities [13–16]. This scientific progress contributed to the better understanding of its biology and of potential treatment targets in some entities [14,17]. Furthermore, molecular profiling enables reverse translation approaches to design more accurate preclinical models [18,19]. In glioblastomas, genetic alterations are clustered into three subtypes: classical, proneuronal, and mesenchymal [14,16,20]. Besides genetical alterations, metabolic adaptation and the reprogramming of cancers cells have been defined as an additional hallmark of cancer [21,22]. In fact, immunosuppressive reshaping of the glioma-associated microenvironment by the release of oncometabolites and a distinct landscape of metabolomic alterations in human glioma have highlighted the importance of metabolic fine tuning for human glioma [23,24]. In vivo imaging using [^{18}F]FET-PET is a strong diagnostic tool in neuro-oncology when assessing tumor metabolism on a molecular level and has become a standard tool in clinical neuroimaging [25]. [^{18}F]FET-PET is an amino acid radiotracer that is taken up by proliferating tumor cells and has been associated with L-type amino acid transporters (LAT) [26]. PET/MRI plays an important clinical role in the evaluation of metabolically active regions and thereby in differentiating therapy-associated pseudoprogression from real tumor progression [7,25]. In particular, [^{18}F]FET-PET improves sensitivity, specificity, and accuracy in differentiating glioma from non-neoplastic tissue [27–29]. [^{18}F]FET-PET/MRI has a value in the differentiation between tumor progression and so-called pseudoprogression [30–33]. Furthermore, contrast-enhanced MRI can evaluate blood brain barrier (BBB) permeability with high sensitivity, and pseudoquantification can be performed [34,35]. Moreover, it has been correlated with tumor infiltration and associated with tumor vascularity, as well as progression-free survival [36,37].

Preclinical models recapulating the genetical and metabolic alterations of human gliomas are urgently needed for dissecting potential vulnerabilities during glioma evolution in preclinical in vivo settings [38]. In this regard, somatic gene transfer systems that offer the spatial and temporal control of gene delivery are of high scientific and translational value [39,40]. One retrovirus-based system that displays these characteristics is the replication-competent avian sarcoma-leukosis virus (RCAS)/tumor virus receptor-A (tva) system [41]. As illustrated in Figure 1A, the RCAS virus offers a multicloning site for the insertion of genetic information with a maximum of 2.5 kilobase (kb), which is controlled by the viral LTR promoter [42]. Viral replication solely takes place in cells expressing the tva receptor, which is essential for viral entrance. Unlike mammalian cells, avian cells, such as the chicken fibroblast long-term cell line DF-1, naturally express the tva receptor (Figure 1B).

The establishment and breeding of immunocompetent, transgenic mice expressing the tva receptor under the control of tissue-specific promoters allows the orthotopic induction of diverse tumor entities using a broad range of oncogenic drivers (example shown in glioma setting in Figure 1C). For instance, mouse models using the RCAS-tva delivery system are available for glioma, ependymoma, pancreatic cancer, hepatic cancer, and ovarian cancer research [41–45]. Additionally, rodent crossbreeding offers large genetic combination possibilities. Examples are the implementation of luciferase-dependent in vivo imaging as well as the combination of existing gene editing systems, such as the widely used cre-lox or CRISPR-Cas9 editing systems for inducing chromosomal translocation or gene fusions [42,46–48].

A platelet-derived growth factor B (PDGFB)-driven glioma model using the RCAS-tva system has been used to test a wide range of novel therapeutic targets [49–51]. It comprises genetic features of the proneuronal glioma subtype by combining PDGFB amplification with the lack of cell cycle regulator Cdkn2a (Figure 1) [14,16,46,52,53].

The aim of the study was to investigate (i) the growth dynamics with a special focus on identifying neuroimaging characteristics of the PDGFB-driven glioma mouse model using clinically relevant imaging methods, i.e., preclinical [^{18}F]FET-PET/MRI, and (ii) to study features of the glioma-associated microenvironment by immunohistochemistry.

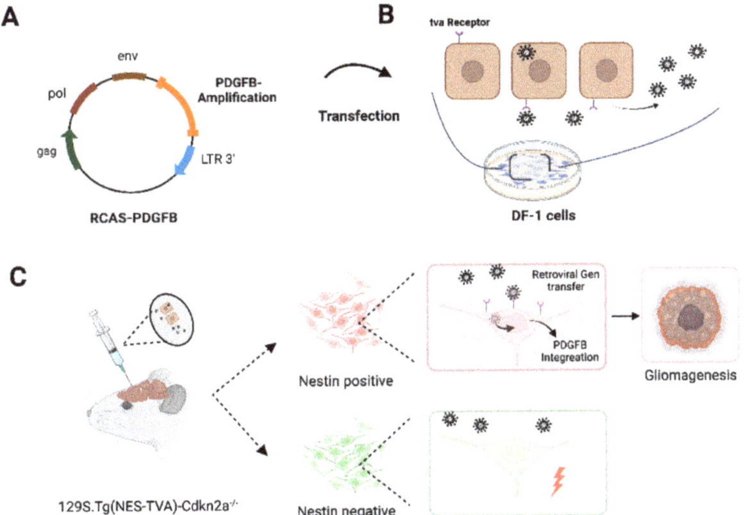

Figure 1. Schematic overview of the PDGFB-driven glioma mouse model using the RCAS-tva delivery system. (**A**) Schematic overview of the RCAS plasmid including a PDGFB amplification. This vector is transfected into DF-1 cells. (**B**) Viral replication takes place after cellular entry via tva receptor binding. (**C**) Implantation of 5×10^4 transfected DF-1 cells into genetic engineered animals. Tissue-dependent expression of the tva receptor is ensured by expression of tva controlled by the nestin promoter in mice. This leads to the infection of only the nestin-positive cell population (red). Subsequently, the PDGFB amplification is integrated retrogradely into the host genome. Together with the systemic deletion of the cell cycle regulator Cdkn2a, intracerebral tumor formation occurs. Created with BioRender.com.

2. Materials and Methods

2.1. DF-1 Cell Transfection

DF-1 cells were cultured in Dulbecco's Modified Eagle's Medium (DMEM) (ThermoFisher, Waltham, MA, USA) at 39 °C with 5% CO_2 atmosphere [54]. Cells were seeded 24 h before transfection. An amount of 2.5 µg of the RCAS-PDGFB or RCAS-GFP plasmid was dissolved in 150 µL DMEM, together with 25 µL SuperFect® Transfection Reagent (Qiagen, Venlo, The Netherlands), and incubated for 7 min at room temperature (RT) to form complexes [41,46]. Next, 1 mL of DMEM was added, and the solution was transferred to the seeded DF-1 cells and incubated for 3 h at 39 °C 5% CO_2. Afterwards, cells were washed and cultured with DMEM. Transfection control was assessed by fluorescence microscopy. The GFP signal of RCAS-GFP-transfected DF-1 cells was evaluated starting at day 5 post transfection. Pictures were taken with an Axiovert 200M imaging system (Zeiss Microscopy, Oberkochen, Germany).

2.2. Transfected DF-1 Cell Implantation into Immunocompetent (129S.Tg(NES-TVA)-Cdkn2a$^{-/-}$) Mice

Animal experiments were conducted in accordance with the local authorities and the German laws regulating the appropriate use of laboratory animals. Described procedures and experimental settings were approved by The Institute of Animal Welfare and the Veterinary Office at the University of Tubingen and the Regional Council Tuebingen. We used a PDGFB-driven glioma mouse model using the RCAS-tva somatic gene transfer delivery system as originally established and described by Hambardzumyan et al. [46]. Fifty thousand RCAS-PDGFB transfected DF-1 cells were implanted intracranially in the same manner as described previously [55–57].

In brief, adult mice (male and female) (129S.Tg(NES-TVA)-Cdkn2a$^{-/-}$) [53] were anesthetized with a 3-component anaesthetic (fentanyl, midazolam, and medetomidine). Then, the anatomical injection site, the right striatum, was located using a stereotactic device (Stoelting, Wood Dale, IL, US). Next, 5×10^4 transfected DF-1 cells resuspended in $1 \times$ PBS, in a volume of 2 µL were injected into the mice using a Hamilton syringe (Hamilton Bonaduz AG, Bonaduz, Switzerland). Intracranially implanted mice were carefully monitored and longitudinally imaged as outlined in Section 2.3. Tumor-bearing mice were euthanized at the appearance of moderate clinical signs, which were regularly assessed according to a predefined, previously described scoring system shown in detail in Supplementary Table S1 [57–60].

2.3. MRI and [^{18}F]FET-PET Measurements

PET and MRI acquisitions were performed longitudinally over a period of 42 days in tumor-bearing mice ($n = 8$), starting 6 days after implantation as outlined in Section 3.1. Isoflurane with a 2.5% induction and 1–1.5% maintenance using room air at a flow rate of 1.5 L/min was used as an anesthetic during the measurements. A constant body temperature of 37 ± 0.5 °C was maintained throughout the acquisitions, using temperature regulated bed systems for MRI (Bruker Biospin, Ettlingen, Germany) and PET (Medres, Cologne, Germany).

MRI scans were performed using a 7T Clinscan small-animal MR scanner equipped with a whole-body transmitter coil and a volume coil that completely covered the mouse head (Bruker Biospin, Ettlingen, Germany). Respiratory rate and gating for MRI sequences were performed using a breathing pad. T2 weighted images (T2W) were acquired using a 2D-spoiled turbo RARE spin echo sequence (256 × 256 matrix, 20 × 20 mm^2 field of view (FOV), repetition time (TR) = 2500 ms, echo time (TE) = 33 ms, slice thickness = 0.7 mm, 18 slices, averages = 2). T1-weighted images were acquired with increasing flip angles using the following parameters: 129 × 129 matrix, 25 × 25 mm^2 field of view (FOV), repetition time (TR) = 10 ms, echo time (TE) = 1.34 ms, slice thickness = 0.2 mm, flip angles = [2,9,27], 80 slices, averages = 2. For contrast enhancement, T1-weighted images were measured using the above-mentioned parameters approximately 4 min after the intravenous contrast agent injection of Gadobutrol (Gadovist$^{®}$ 1 mmol/mL, Bayer Schering Pharma, Berlin, Germany) diluted to a concentration of 0.2 mmol/mL at a dosage of 25 mmol per kg of body weight. Relaxometry T1-maps were calculated by linearly fitting the T1-weighted images voxel-wise, as previously described using Matlab (Matlab 2013b, The MathWorks, Natick, MA, USA) [34,61]. To further validate the T1-maps and quantify gadolinium concentrations in the brain, we acquired T1-weighted images of a phantom containing linearly increasing concentrations of Gadobutrol. PET acquisitions were performed using brain-dedicated beds using an Inveon small-animal PET scanner (Siemens Healthineers, Erlangen, Germany). A bolus of 11.6 ± 0.74 MBq of O-(2-[^{18}F]fluoroethyl)-L-tyrosine ([^{18}F]FET) was injected in the catheterized tail vein and measured using PET through a dynamic acquisition of 50-min in list-mode. The data were histogrammed in a 10-min time frame and reconstructed using an iterative ordered subset expectation maximization (OSEM3D) algorithm. For attenuation correction, a 10-min transmission acquisition using a 57Co source was acquired for every PET measurement.

2.4. MR Image Analysis

The acquired PET and MRI images were realigned and co-registered using Pmod Software v3.2 (Bruker Biospin, Ettlingen, Germany). Regions of interest (ROI) were drawn in agreement with two experienced neuroimaging scientists using the contrast enhanced T1-weighted images. The ROI masks were used to extract data from the T1-relaxometry maps and dynamic PET image data.

2.5. Scoring of Experimental Animals and SMA560&VM/Dk Mice

After transfected DF-1 cell implantation, the animals were closely examined, and the clinical and neurological symptoms were evaluated according to a well-established scoring scheme (Supplementary Table S1) [58]. The endpoint of the experiments was set at a moderate physical burden. As soon as clinical endpoints were met, the experimental animals were euthanized as defined by the responsible governmental authority (Regional council Tuebingen).

Described IHC images in Section 3.3 were taken from either transgenic mice after DF-1 cell implantation or immunocompetent mice of the orthotopic SMA560/VM-Dk model [62]. In brief, 5×10^3 SMA560 cells were implanted in the same manner as described in Section 2.2. Mice were part of an experiment previously published [57]. Shown mice belonged to the control group, which received an isotype control antibody (MOPC-21) intraperitoneally once per week (30 mg/kg). The isotype control was provided by Roche Diagnostics (Penzberg, Germany). Treatment started at day 7 post-tumor cell implantation. The experiment was closed at day 18 post-tumor cell implantation.

2.6. Immunohistochemistry of Murine Tumor Samples

The following antibodies were used: CD3, CD4, CD8, CD11b, CD19, CD20, CD45R, CD163, Ki67, NCRI (Abcam, Cambridge, UK), CD204 (ThermoFisher, Waltham, MA, USA), CD31 (BD Biosciences, Heidelberg, Germany), PD1, and PD-L1 (ProSci, Poway, CA, USA). After reaching the predefined experimental endpoint, animals were perfused with ice cold PBS, and brains were snapped frozen. Then, 8 µm thick sections were cut using a Leica CM3050S cryostat (Leica, Wetzler, Germany). Brain slices were stored at $-80\,°C$.

First, brain sections were dried at room temperature for 10 min, and fixation was performed either with ice-cold acetone at $-20\,°C$ for 10 min or 4% PFA for 15 min. Endogenous peroxidase activity was blocked by Bloxall (Vector Laboratories, Peterborough, UK). Next, brain sections were incubated with 10% bovine serum albumin (BSA) in PBS-Tween 0.3% for 1 h at RT. The primary antibody was incubated overnight at 4 °C. After several washing steps with PBS, slides were incubated for 1 h at RT, with respective biotinylated secondary antibodies diluted in 2% BSA PBS-Tween 0.05%. Vectastain® ABC Kit as a signal amplifier and Vector NovaRED (Vector Laboratories) as a detection kit were used. Counterstaining with Hematoxilin (Sigma-Aldrich, St. Louis, MO, USA) was performed. Finally, slides were dehydrated and were mounted in DPX medium (VWR, Radnor, PA, USA). Stained tissue sections were analysed under a Carl Zeiss Axioplan2 Imaging brightfield microscope (Zeiss Microscopy, Oberkochen, Germany) with the Axio Vision 4.0 software.

2.7. Statistics

In vivo symptom-free survival was evaluated with Kaplan–Meier survival fractions, p values were generated, and the Log-rank test (Mantel–Cox) was performed. Additionally, the Tukey-Kramer post hoc test was used. Error bars represent standard deviation (SD). Pearson correlations were calculated using MATLAB (R2021b, The MathWorks, Inc., Natick, MA, USA).

Growth curves of contrast-enhancing regions and gadolinium- and FET-uptake were analysed by GraphPad Prism 9 (GraphPad Software, San Diego, CA, USA) and visualized with Adobe Illustrator 2022 (Adobe, San José, CA, USA).

3. Results

3.1. Implantation of RCAS-PDGFB Transfected DF-1 Cells and Glioma Formation In Vivo

First, we generated murine glioma in vivo after the implantation of RCAS-PDGFB transfected DF-1 cells in experimental mice in two independent animal experiments with and without longitudinal imaging. Therefore, as illustrated in Figures 1 and 2A, DF-1 cells were transfected with RCAS-PDGFB [46]. Transfection was controlled by a GFP-labelled transfection control (RCAS-GFP), showing an enhanced GFP signal after plasmid transfection (see Figure 2A). Next, 5×10^4 transfected DF-1 cells were implanted into the right striatum of the mice (day 0). For longitudinal MRI and FET-PET imaging, mice received baseline measurement one day before cell implantation (see Figure 2B). Animals received cerebral MRI measurements twice per week (in total, 12 measurements) and an additional FET-PET once per week depending on the availability of the FET tracer, as visualized in Figure 2B. Untreated symptom-free survival was assessed according to the experimental endpoint, as outlined in material/methods and Supplementary Table S1. Median symptom-free survival time in the imaging animal group was 38 days; the independent implantation of transfected DF-1 cells in seven additional animals revealed a median symptom-free survival of 39 days (see Figure 3A).

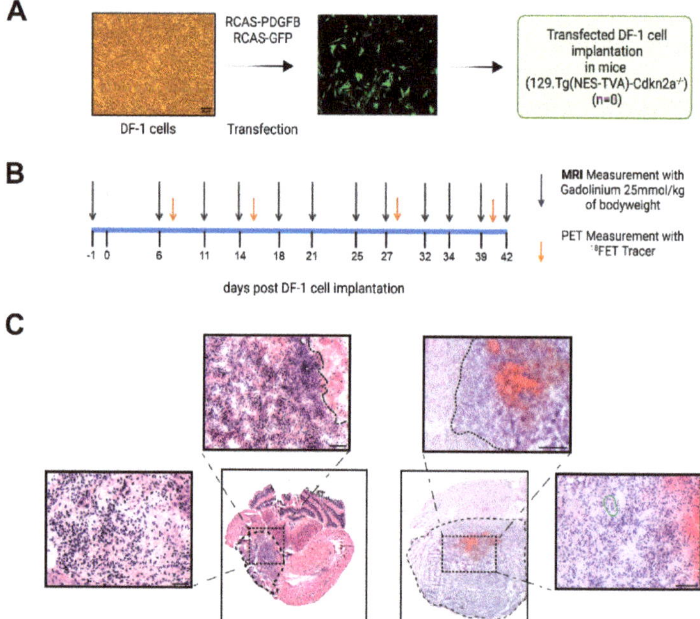

Figure 2. Longitudinal MRI-FET-PET imaging study and H&E staining pattern. (**A**) Schematic overview of experimental design. Brightfield image shows transfected DF-1 cells. Scale bar is 50 μm. Representative picture of DF-1 cells transfected with RCAS-GFP plasmid as transfection control. (**B**) Imaging schedule: day 0 represents the day of implantation of transfected DF-1 cells. (**C**) Representative images of two animals showing glioma formation. Typical aspects of high-grade glioma, such as infiltrating growth behavior, neovascularization (highlighted in green), and cell atypia, are visible. Boundaries of the tumor core area are highlighted with a dotted line. Scale bars are 50 and 100 μm. (**A,B**) were created with BioRender.com.

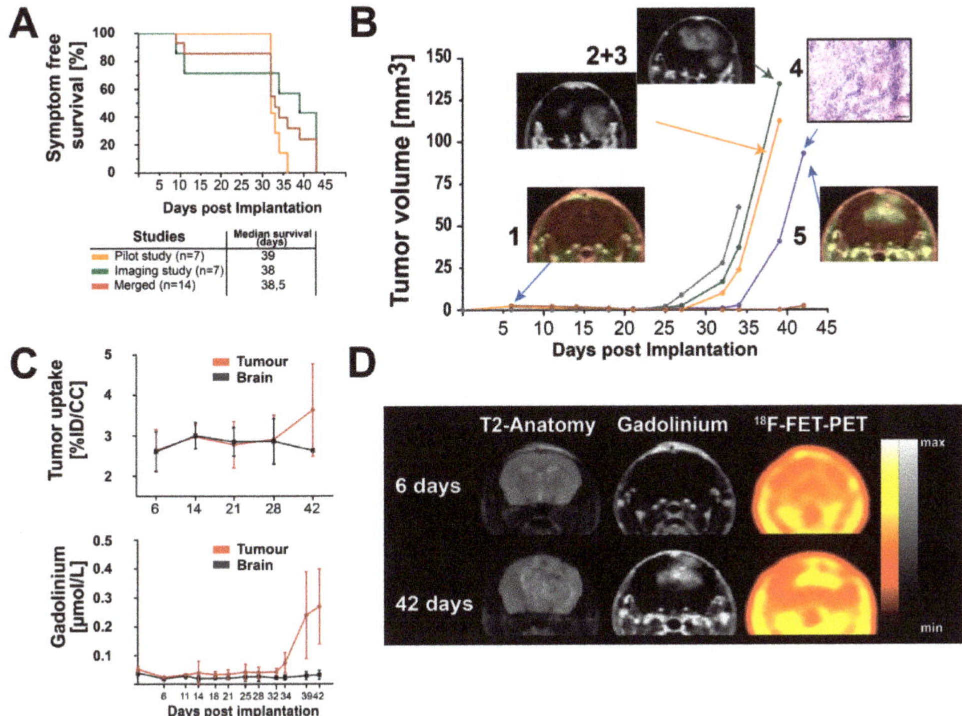

Figure 3. (**A**): Kaplan–Meier plots of symptom-free survival of the longitudinal observational study with and without MRI and FET-PET measurements (yellow and green curve). The red curve represents entire animal studies. No statistically significant differences in Log-Rank test and Tukey–Kramer post hoc test were observed. (**B**) Longitudinal tumor volume development: (1) representative FET-PET image day 7; (2) and (3) representative MRI images of day 39 PTI (post tumor initiation) of two animals showing glioma-like gadolinium enhancement in the final stage of the disease; (4) representative H&E staining showing glioma-like histologic features in the region of the previously occurred gadolinium enhancement (scale bar is 50 µm); (5) representative FET-PET image, showing FET uptake in tumor regions at day 42 PTI. (**C**) (1) Analysis of [^{18}F]FET uptake in five representative imaged animals. (2) Analysis of longitudinal gadolinium uptake in tumor regions and normal brain regions. Error bars represent standard deviation (SD). (**D**): Longitudinal MRI and FET-PET images of one representative animal. Color bars as the reference of gadolinium and [^{18}F]FET uptake are shown.

We performed Hematoxylin and Eosin (H&E) staining for the histological confirmation of tumor formation in post-experimental brain tissue. Additionally, immunohistochemistry for the proliferation marker Ki67 as well as endothelial marker CD31 were executed (see Figure 4, column 1–2). Both experiments revealed extensive tumor formation with glioma-type histological characteristics [20]. We observed infiltrative growth behavior, a high density of strong basophilic stained tumor cells, a high grade of neovascularization, and an extended necrosis area (as illustrated in Figure 2C). This infiltrative growth was more pronounced compared to the SMA560-glioma (see Supplementary Figure S1). Proliferation marker Ki67 was largely present in tumor regions and was good comparable to the staining patterns in untreated VM-Dk mice (see Figure 4, column 1). Staining against CD31 showed large vessel formation in good accordance with the PDGFB amplification as the tumor driver mutation (see Figure 4, column 2). CD31 signal was present in the widely used orthotopic SMA560/VM-Dk glioma mouse model as well. Interestingly, tumor volume expanded over the adjacent brain regions, infiltrating into the adjacent temporal and parietal lobe. Taken together, we could detect robust tumor formation, which could be confirmed by histology, displaying high-grade glioma-like histological features.

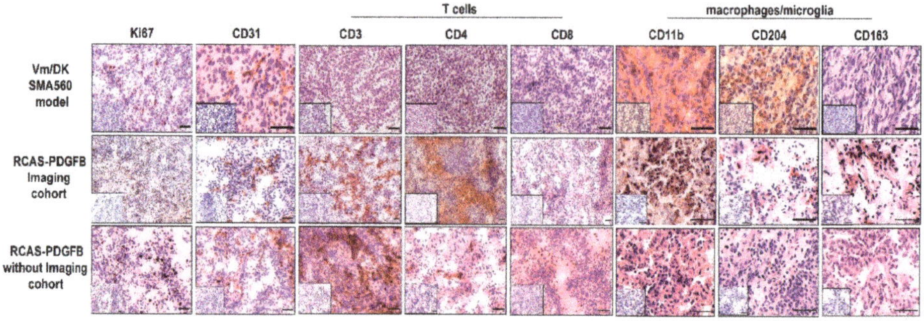

Figure 4. Immunohistochemical comparison of established T cell, microglia/macrophages, and vascularisation markers as indicated above. Representative IHC staining patterns of tumor tissues of either VM/Dk mice treated with an isotype control antibody (MOPC-21), as described in Przystal et al., or transgenic mice after the implantation of RCAS-PDGFB transfected DF1 cells. Small inserts show staining control without the application of primary antibody. Scale bars are 100 µm.

3.2. Longitudinal MR-Imaging Reveals Exponential Growth Dynamics and Late FET Uptake

Acquired imaging data, a total of 12 MRI measurements and 4 FET-PET measurements, were co-registered, and the longitudinal course was analysed. As shown in Figure 3B, the longitudinal volumetry of the contrast agent-enhanced regions revealed tumor-specific signals starting at day 25 until 27 after the implantation of transfected DF-1 cells. First, the contrast agent-enhancing lesions were detected in the region of cell implantation and were expanded and showed increasing accumulations in an exponential fashion. Irregular and diffuse gadolinium uptake, similar to the radiological behavior of human high-grade glioma, was detected. At the final phase of acquired measurements, gadolinium uptake was detected even contralateral to the implantation side, infiltrating the corpus callosum and the contralateral parietal lobe (see Figure 3B(2 + 3),D). In the T2-weighted images, a large signal alteration was apparent in comparable localizations. Furthermore, in the region of the cell implantation, T2-alterations highly suspected for tumor necrosis were detected (Figure 3D). The final measured volume of the contrast agent-enhancing regions before reaching the defined experimental endpoints (as outlined in Supplementary Table S1) were in the range of 65 mm^3 to 135 mm^3 in four of the five included animals (see Figure 3B). Of note, calculated tumor volumes represented up to 40% of the whole brain volume in the final stage of the disease (see Supplementary Figure S2A). As visualized in Figure 3D, [^{18}F]FET uptake was in good accordance with gadolinium uptake in the late phase of tumor progression at day 42. Of note, at 27 days, although several mice presented contrast-enhancing tumor lesions, [^{18}F]FET-PET did not present any focal lesions.

We aimed at investigating the relationships between the measured imaging variables and the timepoints when the primary outcome parameter, i.e., symptom-free survival, was reached. Therefore, we correlated tumor volume, tumor [^{18}F]FET uptake, and gadolinium concentration in the tumor region and the brain parenchyma adjacent to the main tumoral mass. We observed significant positive correlations between tumor volume, [^{18}F]FET uptake, and gadolinium concentrations in the tumor region ($p < 0.001$). Moreover, evaluated imaging variables highly correlated with the primary outcome parameter ($p < 0.001$). Correlations are shown in supplementary Figure S3.

The dynamic analysis of the PET data per time point showed a steady retention in the brain up to day 27 but no tumor differentiation. Unfortunately, we could not perform FET-PET acquisitions between days 27 and 42. After 42 days, the tumors displayed an increased uptake in comparison to brain background, without wash-out dynamics, only constant increased retention (Figure 3C). In summary, the tumors were first identifiable on day 25 using gadolinium-enhanced MRI. We detected colocalizing gadolinium and [^{18}F]FET enhancement on the last imaging acquisition day (day 42 after DF-1 cell implantation).

3.3. PDGFB-Driven Glioma Show High Basal Infiltration of Immune Cells in Comparison with the Orthotopic Syngeneic SMA560/VM-Dk Glioma Mouse Model

Treatment-naïve human glioblastomas harbor an immunosuppressive microenvironment with low numbers of infiltrating T cells and tremendous amounts of tumor-associated macrophages (TAMs) [63,64]. Current approaches in immunotherapy, e.g., by personalized peptide vaccination, can lead to an increased infiltration of CD8-positive T cells and increased immunogenicity [65]. Moreover, systemic immunosuppressive and inflammation markers can be influenced by novel therapeutic options as well [66–69].

Therefore, we assessed the composition of the glioma-associated microenvironment of treatment-naïve PDGFB-driven glioma in mice. We performed an immunohistochemical analysis of 14 markers on tumor tissue (Figure 4 and Supplementary Figure S4). We investigated the infiltration of tumor tissue by host T cells and microglia/macrophages using the well-established markers CD3, CD4, CD8 (as illustrated in Figure 4, column 3–5), CD11b, CD204, and CD163 (column 6–8).

Moreover, we compared the staining patterns with the untreated tissue of the orthotopic SMA560/VM-Dk glioma mouse model (see Figure 4, row 1) [57]. T cell-specific markers showed a stable presence in animal cohorts (both with and without imaging), and the strongest signal was detectable against CD3-positive cells. Of note, even CD8-positive cells could be often clearly distinguished, whereas in the SMA560/VM-DK, only single positive stained cells could be observed (see Figure 4, column 5). Interestingly, CD11b showed a strong staining signal in both glioma mouse models. In contrast CD204, which was widely expressed in the syngeneic SMA560/VM-Dk model, showed a reduced frequency in the evaluated PDGFB-driven glioma. Inversely, we detected an increased CD163 staining signal in the evaluated RCAS-tva model in comparison to the SMA560/VM-Dk model. Additionally, the PD1/PD-L1 axis immunosuppressive markers were assessed (as illustrated in Supplementary Figure S4). PD1 as well as ligand PD-L1 were present in evaluated tissue samples. Additional histological markers against several subtypes of B and T cells revealed comparable results to the SMA560/VM-Dk model, with a slight tendency towards decreased numbers of natural cytotoxicity, triggering receptor 1 (NCRI)-expressing cells (Supplementary Figure S4, column 3) and an increased signal for CD20-expressing cells in the evaluated tumor tissue from mice after DF-1 cell implantation (Supplementary Figure S4, column 2).

4. Discussion

Robust and flexible systems to precisely implement genetic and metabolic alterations in immunocompetent rodent glioma models are essential for the preclinical evaluation of therapeutic strategies [38]. Here, we aimed at further characterizing the PDGFB-driven glioma model using the RCAS-tva delivery system, monitoring tumor growth dynamics as well as tumoral metabolic activity by neuroimaging.

Genetic alterations, such as PDGFB amplification, the homozygous loss of Cyclin Dependent Kinase Inhibitor 2A (CDKN2A), and coding for p16INK4A and p14ARF, show high frequency in human glioma [4]. In general, as outlined in Figures 1 and 2A, we observed robust tumor induction in a reliable manner [41,42], i.e., stable tumor formation and comparable median symptom-free survival, in two independently conducted experiments using the RCAS-tva system with a PDGFB overexpression in Cdkn2a deleted mice, underlining the high reproducibility of the model (Figure 3A). Of note, our evaluated median symptom-free survival was comparable to Hambardzumyan et al., who primarily evaluated the tumor-formation capacity of RCAS-PDGFB in transgenic mice depending on the cerebral implantation site [46]. Control groups of treatment studies, e.g., focusing on colony stimulating factor 1 (CSF1R) inhibition, have also reported similar symptom-free survival [49]. The morphological analysis of post-mortem tumor tissue revealed histological features of human glioblastoma (Figure 2C) [20]. We detected a high grade of neovascularization and necrosis areas as well as infiltrating growth behavior (Figure 2C)). Similar findings were described by Connolly et al., who established a PDGFB-driven glioma model in transgenic rats using the RCAS/tva system [70].

The multimodal neuroradiological assessment of treatment responses using classical and functional imaging modalities, evaluating tumoral growth kinetics, are essential for brain tumor surveillance [7]. Executed longitudinal MRI imaging of the PDGFB-driven glioma model revealed exponential growth kinetics of contrast-enhanced lesions (Figure 3B). Interestingly, the evaluated tumors showed comparable radiological features with the clinical situation. The time points of the first-detectable contrast-enhancing lesions and tumor volumes before reaching the experimental endpoint differed in the range of five to ten days between animals within the experimental group (Figure 3B). Despite the different onset of detectable tumoral lesions, we could observe similar growth patterns and stable median-symptom-free survival, as shown in Figure 3A in our experimental group. These findings correlate with other preclinical studies that evaluated treatment responses by MRI. The exponential growth dynamics were displayed in control groups or in tumors that acquired therapeutic resistance [49,50]. Moreover, growth dynamics might recapitulate high proliferative capacities in combination with rapid neovascularization in human glioblastoma [20].

Due to the complexity of the contrast agent application in rodent models, monitoring glioma growth in therapy studies is mostly performed by assessing T2-weighted images. However, these do not always allow reliable and clear tumor delimitations (Figure 3D). T2-weighted imaging showed perifocal edema with loss of signal in center regions, often observed in cerebral micro bleedings [71]. Of note, tumor borders are rather ill-defined in T2-weighted images (Figure 3B,D). Gadolinium uptake provides a better delineation of the tumor region, similar to the clinical situation in glioma patients [72]. In fact, the observed gadolinium uptake in this study reflected other published contrast-agent uptake patterns observed in the same model during treatment that targeted myeloid-derived suppressor cells (MDSCs) [73]. Interestingly, our findings indicated a delayed contrast-agent uptake starting at day 25 until 27 post-DF-1 cell implantation (Figure 3B). This finding entails a disruption of the BBB that produced a focal observable lesion at day 25. Although, histologically, tumor cells are present at early time points, they appear to present insufficient tissue alterations to disrupt the BBB. A late onset of BBB disruption and gadolinium-enhanced tumor visualization starting between days 25 and 27 emphasizes that appropriate therapy starting points in treatment studies are imperative for evaluating PDGFB-driven glioma. Fixed and early treatment schedules might result in the early therapy initiation of small-size lesions that poorly recapitulate the clinical situation. Accordingly, it might lead to the exaggerated interpretation of treatment responses in rodent models. Therefore, imaging-based therapy start points referring to a predefined minimum tumor volume might be helpful to increase the translational impact, i.e., comparability between in vivo modelling and human glioma patients. This consideration was implemented in several studies tackling TAMs-centered therapies. Therapy started only with a tumor volume of 40 mm^3 in T2-weighted images, largely observed between weeks 4 and 5 after the tumor induction of PDGFB-driven glioma [49,50,74]. In some studies, group randomization referring to initially measured tumor volume was possible in preclinical rodent therapy studies, probably leading to better balanced treatment groups [49,73]. We suggest the following experimental sequence for preclinical therapeutic assessments: cell implantation, baseline imaging and determination of tumor volume, start of therapy with comparable tumor volume in all experimental groups, and clinical and imaging-based monitoring.

Longitudinal metabolic imaging using [^{18}F]FET-PET detecting temporal and spatial alterations in human glioma usually encompasses the period from the initial diagnosis of a clinically apparent tumor to its evolution under therapy. In contrast, the used PDGFB-driven glioma model offers the unique possibility to monitor the interval from tumor induction to the establishment of a clinically and biologically advanced tumor by close-meshed [18F]FET PET imaging time points (Figures 2B and 3). Longitudinal [18F] FET-PET imaging showed the late onset of intense and diffuse tracer uptake in tumoral regions above the background brain activity. The tumor [18F]FET uptake presented a similar pattern as observed post-contrast-enhancement T1-weighted images (Figure 3C,D). Taking these two patterns together, the model presents typical human PET/MRI features that are highly relevant for the diagnostic and clinical management of glioma [33]. Interestingly, [18F]FET-PET did not correlate with BBB permeability at early tumor stages (Figure 3C). This finding is in agreement with previously published glioma rat data, where BBB permeability did not always correlate to [18F]FET uptake [75]. Possible explanations for this phenomenon might be a delayed availability of amino-acid transporters in the tumor, the partial volume effect, or sensitivity limitations of PET [76]. In addition, the lack of a "washout" dynamic curve in all evaluated tumors is reminiscent of the IDH-mutant tumor in humans [77]. Therefore, the uptake behavior of wash-out negative glioblastomas and their biological mechanisms can be further investigated using this mouse model.

Next, we investigated the composition of the glioma-associated microenvironment by immunohistochemistry [63]. We detected several subtypes of infiltrating lymphocytes, as well as CD163- and CD11b-positive cells, highly suggestive of the presence of TAMs inside the glioma-associated microenvironment (Figure 4 and Supplementary Figure S4). The presence of stained TAM markers has been linked to a worse prognosis in molecular glioblastoma subtypes, and an increased frequency correlates with the respective WHO grade [64,78,79]. Additionally, the comparison of the widely used syngeneic SMA560/VM-Dk model showed similarities and differences regarding the composition of the glioma-associated microenvironment, indicating a rodent model-specific microenvironmental structure and reflecting the different genetic landscapes and immune escape mechanisms of modelled human gliomas (Figure 4) [57]. However, potential cofounding factors, such as the difference in tumor size or the experimental setup, might be considered as well. In general, glioma models using the RCAS-tva system might better reflect the natural course of disease than chemically induced or spontaneous-occurring syngeneic orthotopic glioma transplantation models, e.g., implanting GL261 cells in C57BL/6 mice, regarding tumor initiation and disease progression [80].

The PDGFB-driven model was widely used in studies with novel TAM-centered thera-peutic options. For example, one study demonstrated that primarily achieved re-education of TAMs through CSF1R inhibition often led to acquired therapeutic resistance via alter-native insulin growth factor 1 (IGF1)/IGFR signaling. Combination therapeutic regimes targeting the acquired resistance mechanism led to a survival benefit, underscoring the PDGFB-driven mouse model applicability towards combination therapeutic regimes [50]. Moreover, valuable insights into the dynamic composition of the tumor microenvironment under therapy and during disease progression could be achieved in a study combining two-photon microscopy and MRI measurements [74]. Even more so, the discrimination of undiscovered potential drivers of gliomagenesis can be addressed by the establishment of a novel genetic forward screen using the retroviral integration capacity of the RCAS virus for detecting potential novel oncogenes in the PDGFB-driven glioma model [81].

Taken together, we provided the multiparametric profiling of a PDGFB-driven glioma mouse model using the RCAS-tva delivery system and demonstrated radiological, his-tological, and metabolic features that are comparable to human high-grade glioma. Still, the small number of imaged animals as well as limited FET tracer availability during day 27 and 42 are a study limitation. Therefore, future close-meshed imaging studies using [18F]FET PET should begin close to day 25 p.i. in order to capture early model-specific tumor features.

5. Conclusions

Our study provided a multilayered profiling of a PDGFB-driven glioma mouse model using the RCAS-tva delivery system and discovered radiological, histological, and metabolic features that are comparable to human high-grade glioma. We conclude that our results further highlighted the translational capacities of this innovative preclinical model by reflecting relevant glioblastoma-like imaging and histological characteristics. Furthermore, future translational studies using this preclinical model might be further facilitated and reproducible by using the following experimental sequence: cell implantation, baseline imaging and determination of tumor volume, start of therapy with comparable tumor volume in all experimental groups, and clinical and imaging-based monitoring. This might optimize the design of future preclinical studies and comparability with the clinical setting.

Supplementary Materials: The following supporting information can be downloaded at: https://www.mdpi.com/article/10.3390/brainsci12111426/s1, Figure S1: Representative H&E images of SMA560 glioma, Figure S2: Selected immunohistochemistry staining in SMA560 and PDGFB-driven glioma, Figure S3: Longitudinal MRI dynamics and correlation with [^{18}F]FET uptake, Figure S4: Correlation matrix of all imaging parameters to the primary outcome parameter, Table S1: Parameter for scoring of the experimental animals.

Author Contributions: Conceptualization: H.B., S.C.-V. and G.T.; design of animal experiments: H.B., J.M.P., S.C.-V., S.C.B. and G.T.; data acquisition, formal analysis, investigation, and data curation: H.B., K.P., S.C.-V., J.M.P., B.W., F.C.M., D.C., E.C.H., M.T., B.P. and C.I.T.; writing—original draft preparation: H.B. and S.C.-V., with G.T. and E.C.H.; writing—review and editing: all authors; visualization: H.B. and S.C.-V.; supervision: E.C.H. and G.T.; project administration: S.C.B. and G.T.; funding acquisition: G.T. All authors have read and agreed to the published version of the manuscript.

Funding: Parts of this research were funded by a research grant from German Scholars Organisation (GSO/EKFS05), by the Else Kröner Forschungskolleg Tübingen (2019_Kolleg_14), and by the Sigmund-Kiener-Stipendium to H.B. We acknowledge support by the Open Access Publishing Fund of the University of Tübingen.

Institutional Review Board Statement: The animal study protocol was approved by the Institutional Review Board (Regierungspräsidium Tübingen, Germany). The protocol code was N8/15 approved 7 October 2015.

Data Availability Statement: The data presented in this study are available in this article.

Acknowledgments: We thank Heike Pfrommer, Sarah Hendel, and Yeliz Donat for their excellent technical assistance.

Conflicts of Interest: H.B., S.C.-V., J.M.P., D.C., N.K., J.S., M.T., S.C.B. and E.C.H. declare no conflicts of interest. G.T. reports personal fees from BMS, personal fees from AbbVie, personal fees from Novocure, personal fees from Medac, personal fees from Bayer, grants from BMS, grants from Novocure, grants from Roche Diagnostics, and grants from Medac outside the submitted work. The funders had no role in the design of the study; in the collection, analyses, or interpretation of data; in the writing of the manuscript; or in the decision to publish the results.

References

1. Stupp, R.; Mason, W.P.; van den Bent, M.J.; Weller, M.; Fisher, B.; Taphoorn, M.J.; Belanger, K.; Brandes, A.A.; Marosi, C.; Bogdahn, U.; et al. Radiotherapy plus concomitant and adjuvant temozolomide for glioblastoma. *N. Engl. J. Med.* **2005**, *352*, 987–996. [CrossRef] [PubMed]
2. Gilbert, M.R.; Wang, M.; Aldape, K.D.; Stupp, R.; Hegi, M.E.; Jaeckle, K.A.; Armstrong, T.S.; Wefel, J.S.; Won, M.; Blumenthal, D.T.; et al. Dose-dense temozolomide for newly diagnosed glioblastoma: A randomized phase III clinical trial. *J. Clin. Oncol.* **2013**, *31*, 4085–4091. [CrossRef] [PubMed]
3. Stupp, R.; Taillibert, S.; Kanner, A.; Read, W.; Steinberg, D.; Lhermitte, B.; Toms, S.; Idbaih, A.; Ahluwalia, M.S.; Fink, K.; et al. Effect of Tumor-Treating Fields Plus Maintenance Temozolomide vs Maintenance Temozolomide Alone on Survival in Patients With Glioblastoma: A Randomized Clinical Trial. *JAMA* **2017**, *318*, 2306–2316. [CrossRef] [PubMed]
4. Chinot, O.L.; Wick, W.; Mason, W.; Henriksson, R.; Saran, F.; Nishikawa, R.; Carpentier, A.F.; Hoang-Xuan, K.; Kavan, P.; Cernea, D.; et al. Bevacizumab plus radiotherapy-temozolomide for newly diagnosed glioblastoma. *N. Engl. J. Med.* **2014**, *370*, 709–722. [CrossRef] [PubMed]
5. Gilbert, M.R.; Dignam, J.J.; Armstrong, T.S.; Wefel, J.S.; Blumenthal, D.T.; Vogelbaum, M.A.; Colman, H.; Chakravarti, A.; Pugh, S.; Won, M.; et al. A randomized trial of bevacizumab for newly diagnosed glioblastoma. *N. Engl. J. Med.* **2014**, *370*, 699–708. [CrossRef]

6.	Stupp, R.; Hegi, M.E.; Mason, W.P.; van den Bent, M.J.; Taphoorn, M.J.B.; Janzer, R.C.; Ludwin, S.K.; Allgeier, A.; Fisher, B.; Belanger, K.; et al. Effects of radiotherapy with concomitant and adjuvant temozolomide versus radiotherapy alone on survival in glioblastoma in a randomised phase III study: 5-year analysis of the EORTC-NCIC trial. *Lancet Oncol.* **2009**, *10*, 459–466. [CrossRef]

7.	Weller, M.; van den Bent, M.; Preusser, M.; Le Rhun, E.; Tonn, J.C.; Minniti, G.; Bendszus, M.; Balana, C.; Chinot, O.; Dirven, L.; et al. EANO guidelines on the diagnosis and treatment of diffuse gliomas of adulthood. *Nat. Rev. Clin. Oncol.* **2021**, *18*, 170–186. [CrossRef]

8.	Herrlinger, U.; Tzaridis, T.; Mack, F.; Steinbach, J.P.; Schlegel, U.; Sabel, M.; Hau, P.; Kortmann, R.-D.; Krex, D.; Grauer, O.; et al. Lomustine-temozolomide combination therapy versus standard temozolomide therapy in patients with newly diagnosed glioblastoma with methylated MGMT promoter (CeTeG/NOA–09): A randomised, open-label, phase 3 trial. *Lancet* **2019**, *393*, 678–688. [CrossRef]

9.	Suchorska, B.; Weller, M.; Tabatabai, G.; Senft, C.; Hau, P.; Sabel, M.C.; Herrlinger, U.; Ketter, R.; Schlegel, U.; Marosi, C.; et al. Complete resection of contrast-enhancing tumor volume is associated with improved survival in recurrent glioblastoma-results from the DIRECTOR trial. *Neuro Oncol.* **2016**, *18*, 549–556. [CrossRef]

10.	Pasqualetti, F.; Montemurro, N.; Desideri, I.; Loi, M.; Giannini, N.; Gadducci, G.; Malfatti, G.; Cantarella, M.; Gonnelli, A.; Montrone, S.; et al. Impact of recurrence pattern in patients undergoing a second surgery for recurrent glioblastoma. *Acta Neurol. Belg.* **2022**, *122*, 441–446. [CrossRef]

11.	Sahm, F.; Capper, D.; Jeibmann, A.; Habel, A.; Paulus, W.; Troost, D.; von Deimling, A. Addressing diffuse glioma as a systemic brain disease with single-cell analysis. *Arch. Neurol.* **2012**, *69*, 523–526. [CrossRef] [PubMed]

12.	Vogelstein, B.; Papadopoulos, N.; Velculescu, V.E.; Zhou, S.; Diaz, L.A., Jr.; Kinzler, K.W. Cancer genome landscapes. *Science* **2013**, *339*, 1546–1558. [CrossRef]

13.	Kool, M.; Korshunov, A.; Remke, M.; Jones, D.T.; Schlanstein, M.; Northcott, P.A.; Cho, Y.J.; Koster, J.; Schouten-van Meeteren, A.; van Vuurden, D.; et al. Molecular subgroups of medulloblastoma: An international meta-analysis of transcriptome, genetic aberrations, and clinical data of WNT, SHH, Group 3, and Group 4 medulloblastomas. *Acta Neuropathol.* **2012**, *123*, 473–484. [CrossRef] [PubMed]

14.	Wang, Q.; Hu, B.; Hu, X.; Kim, H.; Squatrito, M.; Scarpace, L.; deCarvalho, A.C.; Lyu, S.; Li, P.; Li, Y.; et al. Tumor Evolution of Glioma-Intrinsic Gene Expression Subtypes Associates with Immunological Changes in the Microenvironment. *Cancer Cell* **2017**, *32*, 42–56.e46. [CrossRef] [PubMed]

15.	Brennan, C.W.; Verhaak, R.G.; McKenna, A.; Campos, B.; Noushmehr, H.; Salama, S.R.; Zheng, S.; Chakravarty, D.; Sanborn, J.Z.; Berman, S.H.; et al. The somatic genomic landscape of glioblastoma. *Cell* **2013**, *155*, 462–477. [CrossRef] [PubMed]

16.	Verhaak, R.G.; Hoadley, K.A.; Purdom, E.; Wang, V.; Qi, Y.; Wilkerson, M.D.; Miller, C.R.; Ding, L.; Golub, T.; Mesirov, J.P.; et al. Integrated genomic analysis identifies clinically relevant subtypes of glioblastoma characterized by abnormalities in PDGFRA, IDH1, EGFR, and NF1. *Cancer Cell* **2010**, *17*, 98–110. [CrossRef] [PubMed]

17.	The AACR Project GENIE Consortium ; André, F.; Arnedos, M.; Baras, A.S.; Baselga, J.; Bedard, P.L.; Berger, M.F.; Bierkens, M.; Calvo, F.; Cerami, E.; et al. AACR Project GENIE: Powering Precision Medicine through an International Consortium. *Cancer Discov.* **2017**, *7*, 818–831. [CrossRef]

18.	Honkala, A.; Malhotra, S.V.; Kummar, S.; Junttila, M.R. Harnessing the predictive power of preclinical models for oncology drug development. *Nat. Rev. Drug Discov.* **2022**, *21*, 99–114. [CrossRef]

19.	Cancer models for reverse and forward translation. *Nat. Cancer* **2022**, *3*, 135. [CrossRef]

20.	Louis, D.N.; Perry, A.; Wesseling, P.; Brat, D.J.; Cree, I.A.; Figarella-Branger, D.; Hawkins, C.; Ng, H.K.; Pfister, S.M.; Reifenberger, G.; et al. The 2021 WHO Classification of Tumors of the Central Nervous System: A summary. *Neuro Oncol.* **2021**, *23*, 1231–1251. [CrossRef]

21.	Hanahan, D. Hallmarks of Cancer: New Dimensions. *Cancer Discov.* **2022**, *12*, 31–46. [CrossRef] [PubMed]

22.	Hanahan, D.; Weinberg, R.A. Hallmarks of Cancer: The Next Generation. *Cell* **2011**, *144*, 646–674. [CrossRef] [PubMed]

23.	Trautwein, C.; Zizmare, L.; Maurer, I.; Bender, B.; Bayer, B.; Ernemann, U.; Tatagiba, M.; Grau, S.J.; Pichler, B.J.; Skardelly, M.; et al. Tissue metabolites in diffuse glioma and their modulations by IDH1 mutation, histology, and treatment. *JCI Insight* **2022**, *7*, e153526. [CrossRef] [PubMed]

24.	Bunse, L.; Pusch, S.; Bunse, T.; Sahm, F.; Sanghvi, K.; Friedrich, M.; Alansary, D.; Sonner, J.K.; Green, E.; Deumelandt, K.; et al. Suppression of antitumor T cell immunity by the oncometabolite (R)-2-hydroxyglutarate. *Nat. Med.* **2018**, *24*, 1192–1203. [CrossRef] [PubMed]

25.	Law, I.; Albert, N.L.; Arbizu, J.; Boellaard, R.; Drzezga, A.; Galldiks, N.; la Fougere, C.; Langen, K.J.; Lopci, E.; Lowe, V.; et al. Joint EANM/EANO/RANO practice guidelines/SNMMI procedure standards for imaging of gliomas using PET with radiolabelled amino acids and [(18)F]FDG: Version 1.0. *Eur. J. Nucl. Med. Mol. Imaging* **2019**, *46*, 540–557. [CrossRef]

26.	Götz, I.; Grosu, A. [18F]FET-PET Imaging for Treatment and Response Monitoring of Radiation Therapy in Malignant Glioma Patients—A Review. *Front. Oncol.* **2013**, *3*, 104. [CrossRef]

27.	Pauleit, D.; Floeth, F.; Hamacher, K.; Riemenschneider, M.J.; Reifenberger, G.; Müller, H.W.; Zilles, K.; Coenen, H.H.; Langen, K.J. O-(2-[18F]fluoroethyl)-L-tyrosine PET combined with MRI improves the diagnostic assessment of cerebral gliomas. *Brain* **2005**, *128*, 678–687. [CrossRef]

28. Galldiks, N.; Dunkl, V.; Stoffels, G.; Hutterer, M.; Rapp, M.; Sabel, M.; Reifenberger, G.; Kebir, S.; Dorn, F.; Blau, T.; et al. Diagnosis of pseudoprogression in patients with glioblastoma using O-(2-[18F]fluoroethyl)-L-tyrosine PET. *Eur. J. Nucl. Med. Mol. Imaging* **2015**, *42*, 685–695. [CrossRef]

29. Floeth, F.W.; Pauleit, D.; Wittsack, H.J.; Langen, K.J.; Reifenberger, G.; Hamacher, K.; Messing-Jünger, M.; Zilles, K.; Weber, F.; Stummer, W.; et al. Multimodal metabolic imaging of cerebral gliomas: Positron emission tomography with [18F]fluoroethyl-L-tyrosine and magnetic resonance spectroscopy. *J. Neurosurg.* **2005**, *102*, 318–327. [CrossRef]

30. Kristin Schmitz, A.; Sorg, R.V.; Stoffels, G.; Grauer, O.M.; Galldiks, N.; Steiger, H.-J.; Kamp, M.A.; Langen, K.-J.; Sabel, M.; Rapp, M. Diagnostic impact of additional O-(2-[18F]fluoroethyl)-L-tyrosine (18F-FET) PET following immunotherapy with dendritic cell vaccination in glioblastoma patients. *Br. J. Neurosurg.* **2021**, *35*, 736–742. [CrossRef]

31. Antonios, J.P.; Soto, H.; Everson, R.G.; Moughon, D.L.; Wang, A.C.; Orpilla, J.; Radu, C.; Ellingson, B.M.; Lee, J.T.; Cloughesy, T.; et al. Detection of immune responses after immunotherapy in glioblastoma using PET and MRI. *Proc. Natl. Acad. Sci.* **2017**, *114*, 10220–10225. [CrossRef] [PubMed]

32. Pyka, T.; Hiob, D.; Preibisch, C.; Gempt, J.; Wiestler, B.; Schlegel, J.; Straube, C.; Zimmer, C. Diagnosis of glioma recurrence using multiparametric dynamic 18F-fluoroethyl-tyrosine PET-MRI. *Eur. J. Radiol.* **2018**, *103*, 32–37. [CrossRef] [PubMed]

33. Brendle, C.; Maier, C.; Bender, B.; Schittenhelm, J.; Paulsen, F.; Renovanz, M.; Roder, C.; Castaneda-Vega, S.; Tabatabai, G.; Ernemann, U.; et al. Impact of ^{18}F-FET PET/MRI on Clinical Management of Brain Tumor Patients. *J. Nucl. Med.* **2022**, *63*, 522–527. [CrossRef] [PubMed]

34. Vega, S.C.; Leiss, V.; Piekorz, R.; Calaminus, C.; Pexa, K.; Vuozzo, M.; Schmid, A.M.; Devanathan, V.; Kesenheimer, C.; Pichler, B.J.; et al. Selective protection of murine cerebral Gi/o-proteins from inactivation by parenterally injected pertussis toxin. *J. Mol. Med.* **2020**, *98*, 97–110. [CrossRef]

35. Weinl, C.; Castaneda Vega, S.; Riehle, H.; Stritt, C.; Calaminus, C.; Wolburg, H.; Mauel, S.; Breithaupt, A.; Gruber, A.D.; Wasylyk, B.; et al. Endothelial depletion of murine SRF/MRTF provokes intracerebral hemorrhagic stroke. *Proc. Natl. Acad. Sci.* **2015**, *112*, 9914–9919. [CrossRef]

36. Kang, Y.; Hong, E.K.; Rhim, J.H.; Yoo, R.E.; Kang, K.M.; Yun, T.J.; Kim, J.H.; Sohn, C.H.; Park, S.W.; Choi, S.H. Prognostic Value of Dynamic Contrast-Enhanced MRI-Derived Pharmacokinetic Variables in Glioblastoma Patients: Analysis of Contrast-Enhancing Lesions and Non-Enhancing T2 High-Signal Intensity Lesions. *Korean J. Radiol.* **2020**, *21*, 707–716. [CrossRef]

37. Muller, A.; Jurcoane, A.; Kebir, S.; Ditter, P.; Schrader, F.; Herrlinger, U.; Tzaridis, T.; Madler, B.; Schild, H.H.; Glas, M.; et al. Quantitative T1-mapping detects cloudy-enhancing tumor compartments predicting outcome of patients with glioblastoma. *Cancer Med.* **2017**, *6*, 89–99. [CrossRef]

38. Aldape, K.; Brindle, K.M.; Chesler, L.; Chopra, R.; Gajjar, A.; Gilbert, M.R.; Gottardo, N.; Gutmann, D.H.; Hargrave, D.; Holland, E.C.; et al. Challenges to curing primary brain tumours. *Nat. Rev. Clin. Oncol.* **2019**, *16*, 509–520. [CrossRef]

39. Lenting, K.; Verhaak, R.; ter Laan, M.; Wesseling, P.; Leenders, W. Glioma: Experimental models and reality. *Acta Neuropathol.* **2017**, *133*, 263–282. [CrossRef]

40. Olson, B.; Li, Y.; Lin, Y.; Liu, E.T.; Patnaik, A. Mouse Models for Cancer Immunotherapy Research. *Cancer Discov.* **2018**, *8*, 1358–1365. [CrossRef]

41. von Werder, A.; Seidler, B.; Schmid, R.M.; Schneider, G.; Saur, D. Production of avian retroviruses and tissue-specific somatic retroviral gene transfer in vivo using the RCAS/TVA system. *Nat. Protoc.* **2012**, *7*, 1167–1183. [CrossRef] [PubMed]

42. Kanvinde, P.P.; Malla, A.P.; Connolly, N.P.; Szulzewsky, F.; Anastasiadis, P.; Ames, H.M.; Kim, A.J.; Winkles, J.A.; Holland, E.C.; Woodworth, G.F. Leveraging the replication-competent avian-like sarcoma virus/tumor virus receptor-A system for modeling human gliomas. *Glia* **2021**, *69*, 2059–2076. [CrossRef] [PubMed]

43. Ozawa, T.; Arora, S.; Szulzewsky, F.; Juric-Sekhar, G.; Miyajima, Y.; Bolouri, H.; Yasui, Y.; Barber, J.; Kupp, R.; Dalton, J.; et al. A De Novo Mouse Model of C11orf95-RELA Fusion-Driven Ependymoma Identifies Driver Functions in Addition to NF-κB. *Cell Rep.* **2018**, *23*, 3787–3797. [CrossRef] [PubMed]

44. Lewis, B.C.; Klimstra, D.S.; Socci, N.D.; Xu, S.; Koutcher, J.A.; Varmus, H.E. The absence of p53 promotes metastasis in a novel somatic mouse model for hepatocellular carcinoma. *Mol. Cell Biol.* **2005**, *25*, 1228–1237. [CrossRef]

45. Yuan, Z.; Gardiner, J.C.; Maggi, E.C.; Adem, A.; Zhang, G.; Lee, S.; Romanienko, P.; Du, Y.-C.N.; Libutti, S.K. Tissue-specific induced DNA methyltransferase 1 (Dnmt1) in endocrine pancreas by RCAS-TVA-based somatic gene transfer system promotes β-cell proliferation. *Cancer Gene Ther.* **2019**, *26*, 94–102. [CrossRef]

46. Hambardzumyan, D.; Amankulor, N.M.; Helmy, K.Y.; Becher, O.J.; Holland, E.C. Modeling Adult Gliomas Using RCAS/t-va Technology. *Transl. Oncol.* **2009**, *2*, 89–IN86. [CrossRef]

47. Uhrbom, L.; Nerio, E.; Holland, E.C. Dissecting tumor maintenance requirements using bioluminescence imaging of cell proliferation in a mouse glioma model. *Nat. Med.* **2004**, *10*, 1257–1260. [CrossRef]

48. Oldrini, B.; Curiel-García, Á.; Marques, C.; Matia, V.; Uluçkan, Ö.; Graña-Castro, O.; Torres-Ruiz, R.; Rodriguez-Perales, S.; Huse, J.T.; Squatrito, M. Somatic genome editing with the RCAS-TVA-CRISPR-Cas9 system for precision tumor modeling. *Nat. Commun.* **2018**, *9*, 1466. [CrossRef]

49. Pyonteck, S.M.; Akkari, L.; Schuhmacher, A.J.; Bowman, R.L.; Sevenich, L.; Quail, D.F.; Olson, O.C.; Quick, M.L.; Huse, J.T.; Teijeiro, V.; et al. CSF-1R inhibition alters macrophage polarization and blocks glioma progression. *Nat. Med.* **2013**, *19*, 1264–1272. [CrossRef]

50. Quail, D.F.; Bowman, R.L.; Akkari, L.; Quick, M.L.; Schuhmacher, A.J.; Huse, J.T.; Holland, E.C.; Sutton, J.C.; Joyce, J.A. The tumor microenvironment underlies acquired resistance to CSF-1R inhibition in gliomas. *Science* **2016**, *352*, aad3018. [CrossRef]

51. Pitter, K.L.; Tamagno, I.; Alikhanyan, K.; Hosni-Ahmed, A.; Pattwell, S.S.; Donnola, S.; Dai, C.; Ozawa, T.; Chang, M.; Chan, T.A.; et al. Corticosteroids compromise survival in glioblastoma. *Brain* **2016**, *139*, 1458–1471. [CrossRef] [PubMed]

52. Tchougounova, E.; Kastemar, M.; Brasater, D.; Holland, E.C.; Westermark, B.; Uhrbom, L. Loss of Arf causes tumor progression of PDGFB-induced oligodendroglioma. *Oncogene* **2007**, *26*, 6289–6296. [CrossRef] [PubMed]

53. Dai, C.; Celestino, J.C.; Okada, Y.; Louis, D.N.; Fuller, G.N.; Holland, E.C. PDGF autocrine stimulation dedifferentiates cultured astrocytes and induces oligodendrogliomas and oligoastrocytomas from neural progenitors and astrocytes in vivo. *Genes Dev.* **2001**, *15*, 1913–1925. [CrossRef] [PubMed]

54. Himly, M.; Foster, D.N.; Bottoli, I.; Iacovoni, J.S.; Vogt, P.K. The DF-1 Chicken Fibroblast Cell Line: Transformation Induced by Diverse Oncogenes and Cell Death Resulting from Infection by Avian Leukosis Viruses. *Virology* **1998**, *248*, 295–304. [CrossRef]

55. Miller, J.; Eisele, G.; Tabatabai, G.; Aulwurm, S.; von Kurthy, G.; Stitz, L.; Roth, P.; Weller, M. Soluble CD70: A novel immunotherapeutic agent for experimental glioblastoma. *J. Neurosurg.* **2010**, *113*, 280–285. [CrossRef]

56. Tabatabai, G.; Hasenbach, K.; Herrmann, C.; Maurer, G.; Mohle, R.; Marini, P.; Grez, M.; Wick, W.; Weller, M. Glioma tropism of lentivirally transduced hematopoietic progenitor cells. *Int. J. Oncol.* **2010**, *36*, 1409–1417.

57. Przystal, J.M.; Becker, H.; Canjuga, D.; Tsiami, F.; Anderle, N.; Keller, A.-L.; Pohl, A.; Ries, C.H.; Schmittnaegel, M.; Korinetska, N.; et al. Targeting CSF1R Alone or in Combination with PD1 in Experimental Glioma. *Cancers* **2021**, *13*, 2400. [CrossRef]

58. Langford, D.J.; Bailey, A.L.; Chanda, M.L.; Clarke, S.E.; Drummond, T.E.; Echols, S.; Glick, S.; Ingrao, J.; Klassen-Ross, T.; Lacroix-Fralish, M.L.; et al. Coding of facial expressions of pain in the laboratory mouse. *Nat. Methods* **2010**, *7*, 447–449. [CrossRef]

59. Walter, B.; Canjuga, D.; Yüz, S.G.; Ghosh, M.; Bozko, P.; Przystal, J.M.; Govindarajan, P.; Anderle, N.; Keller, A.L.; Tatagiba, M.; et al. Argyrin F Treatment—Induced Vulnerabilities Lead to a Novel Combination Therapy in Experimental Glioma. *Adv. Ther.* **2021**, *4*, 2100078. [CrossRef]

60. Koch, M.S.; Czemmel, S.; Lennartz, F.; Beyeler, S.; Rajaraman, S.; Przystal, J.M.; Govindarajan, P.; Canjuga, D.; Neumann, M.; Rizzu, P.; et al. Experimental glioma with high bHLH expression harbor increased replicative stress and are sensitive toward ATR inhibition. *Neuro-Oncol. Adv.* **2020**, *2*, vdaa115. [CrossRef]

61. Michelotti, F.C.; Bowden, G.; Küppers, A.; Joosten, L.; Maczewsky, J.; Nischwitz, V.; Drews, G.; Maurer, A.; Gotthardt, M.; Schmid, A.M.; et al. PET/MRI enables simultaneous in vivo quantification of β-cell mass and function. *Theranostics* **2020**, *10*, 398–410. [CrossRef] [PubMed]

62. Serano, R.D.; Pegram, C.N.; Bigner, D.D. Tumorigenic cell culture lines from a spontaneous VM/Dk murine astrocytoma (SMA). *Acta Neuropathol.* **1980**, *51*, 53–64. [CrossRef] [PubMed]

63. Berghoff, A.S.; Kiesel, B.; Widhalm, G.; Rajky, O.; Ricken, G.; Wohrer, A.; Dieckmann, K.; Filipits, M.; Brandstetter, A.; Weller, M.; et al. Programmed death ligand 1 expression and tumor-infiltrating lymphocytes in glioblastoma. *Neuro Oncol.* **2015**, *17*, 1064–1075. [CrossRef]

64. Quail, D.F.; Joyce, J.A. The Microenvironmental Landscape of Brain Tumors. *Cancer Cell* **2017**, *31*, 326–341. [CrossRef]

65. Hilf, N.; Kuttruff-Coqui, S.; Frenzel, K.; Bukur, V.; Stevanovic, S.; Gouttefangeas, C.; Platten, M.; Tabatabai, G.; Dutoit, V.; van der Burg, S.H.; et al. Actively personalized vaccination trial for newly diagnosed glioblastoma. *Nature* **2019**, *565*, 240–245. [CrossRef] [PubMed]

66. Berghoff, A.S.; Preusser, M. Does neoadjuvant anti-PD1 therapy improve glioblastoma outcome? *Nat. Rev. Neurol.* **2019**, *15*, 314–315. [CrossRef]

67. Cloughesy, T.F.; Mochizuki, A.Y.; Orpilla, J.R.; Hugo, W.; Lee, A.H.; Davidson, T.B.; Wang, A.C.; Ellingson, B.M.; Rytlewski, J.A.; Sanders, C.M.; et al. Neoadjuvant anti-PD-1 immunotherapy promotes a survival benefit with intratumoral and systemic immune responses in recurrent glioblastoma. *Nat. Med.* **2019**, *25*, 477–486. [CrossRef]

68. Schalper, K.A.; Rodriguez-Ruiz, M.E.; Diez-Valle, R.; Lopez-Janeiro, A.; Porciuncula, A.; Idoate, M.A.; Inoges, S.; de Andrea, C.; Lopez-Diaz de Cerio, A.; Tejada, S.; et al. Neoadjuvant nivolumab modifies the tumor immune microenvironment in resectable glioblastoma. *Nat. Med.* **2019**, *25*, 470–476. [CrossRef]

69. Pasqualetti, F.; Giampietro, C.; Montemurro, N.; Giannini, N.; Gadducci, G.; Orlandi, P.; Natali, E.; Chiarugi, P.; Gonnelli, A.; Cantarella, M.; et al. Old and New Systemic Immune-Inflammation Indexes Are Associated with Overall Survival of Glioblastoma Patients Treated with Radio-Chemotherapy. *Genes* **2022**, *13*, 1054. [CrossRef]

70. Connolly, N.P.; Stokum, J.A.; Schneider, C.S.; Ozawa, T.; Xu, S.; Galisteo, R.; Castellani, R.J.; Kim, A.J.; Simard, J.M.; Winkles, J.A.; et al. Genetically engineered rat gliomas: PDGF-driven tumor initiation and progression in tv-a transgenic rats recreate key features of human brain cancer. *PLoS ONE* **2017**, *12*, e0174557. [CrossRef]

71. Castaneda Vega, S.; Weinl, C.; Calaminus, C.; Wang, L.; Harant, M.; Ehrlichmann, W.; Thiele, D.; Kohlhofer, U.; Reischl, G.; Hempel, J.-M.; et al. Characterization of a novel murine model for spontaneous hemorrhagic stroke using in vivo PET and MR multiparametric imaging. *NeuroImage* **2017**, *155*, 245–256. [CrossRef] [PubMed]

72. Oh, J.; Cha, S.; Aiken, A.H.; Han, E.T.; Crane, J.C.; Stainsby, J.A.; Wright, G.A.; Dillon, W.P.; Nelson, S.J. Quantitative apparent diffusion coefficients and T2 relaxation times in characterizing contrast enhancing brain tumors and regions of peritumoral edema. *J. Magn. Reson. Imaging* **2005**, *21*, 701–708. [CrossRef] [PubMed]

73. Raychaudhuri, B.; Rayman, P.; Huang, P.; Grabowski, M.; Hambardzumyan, D.; Finke, J.H.; Vogelbaum, M.A. Myeloid derived suppressor cell infiltration of murine and human gliomas is associated with reduction of tumor infiltrating lymphocytes. *J Neurooncol.* **2015**, *122*, 293–301. [CrossRef] [PubMed]

74. Zomer, A.; Croci, D.; Kowal, J.; van Gurp, L.; Joyce, J.A. Multimodal imaging of the dynamic brain tumor microenvironment during glioblastoma progression and in response to treatment. *iScience* **2022**, *25*, 104570. [CrossRef]

75. Stegmayr, C.; Bandelow, U.; Oliveira, D.; Lohmann, P.; Willuweit, A.; Filss, C.; Galldiks, N.; Lübke, J.H.R.; Shah, N.J.; Ermert, J.; et al. Influence of blood-brain barrier permeability on O-(2-18F-fluoroethyl)-L-tyrosine uptake in rat gliomas. *Eur. J. Nucl. Med. Mol. Imaging* **2017**, *44*, 408–416. [CrossRef]

76. Liesche, F.; Lukas, M.; Preibisch, C.; Shi, K.; Schlegel, J.; Meyer, B.; Schwaiger, M.; Zimmer, C.; Förster, S.; Gempt, J.; et al. 18F-Fluoroethyl-tyrosine uptake is correlated with amino acid transport and neovascularization in treatment-naive glioblastomas. *Eur. J. Nucl. Med. Mol. Imaging* **2019**, *46*, 2163–2168. [CrossRef]

77. Kunz, M.; Albert, N.L.; Unterrainer, M.; la Fougere, C.; Egensperger, R.; Schüller, U.; Lutz, J.; Kreth, S.; Tonn, J.-C.; Kreth, F.-W.; et al. Dynamic 18F-FET PET is a powerful imaging biomarker in gadolinium-negative gliomas. *Neuro-Oncol.* **2018**, *21*, 274–284. [CrossRef]

78. Komohara, Y.; Ohnishi, K.; Kuratsu, J.; Takeya, M. Possible involvement of the M2 anti-inflammatory macrophage phenotype in growth of human gliomas. *J. Pathol.* **2008**, *216*, 15–24. [CrossRef]

79. Martinez-Lage, M.; Lynch, T.M.; Bi, Y.; Cocito, C.; Way, G.P.; Pal, S.; Haller, J.; Yan, R.E.; Ziober, A.; Nguyen, A.; et al. Immune landscapes associated with different glioblastoma molecular subtypes. *Acta Neuropathol. Commun.* **2019**, *7*, 203. [CrossRef]

80. Oh, T.; Fakurnejad, S.; Sayegh, E.T.; Clark, A.J.; Ivan, M.E.; Sun, M.Z.; Safaee, M.; Bloch, O.; James, C.D.; Parsa, A.T. Immunocompetent murine models for the study of glioblastoma immunotherapy. *J. Transl. Med.* **2014**, *12*, 107. [CrossRef]

81. Weishaupt, H.; Čančer, M.; Rosén, G.; Holmberg, K.O.; Häggqvist, S.; Bunikis, I.; Jiang, Y.; Sreedharan, S.; Gyllensten, U.; Becher, O.J.; et al. Novel cancer gene discovery using a forward genetic screen in RCAS-PDGFB-driven gliomas. *Neuro Oncol.* **2022**, noac158. [CrossRef] [PubMed]

Article

A Novel Necroptosis-Related Prognostic Signature of Glioblastoma Based on Transcriptomics Analysis and Single Cell Sequencing Analysis

Yiwen Wu [1], Yi Huang [1,2,3,*], Chenhui Zhou [1,2,3], Haifeng Wang [1], Zhepei Wang [1,2], Jiawei Wu [1], Sheng Nie [1], Xinpeng Deng [1,2], Jie Sun [1] and Xiang Gao [1,*]

[1] Department of Neurosurgery, Ningbo First Hospital, Zhejiang University School of Medicine, Ningbo 315010, China; wuyiwen@zju.edu.cn (Y.W.); zhouchenhui1989@126.com (C.Z.); 18857496593@139.com (H.W.); fagoce@126.com (Z.W.); 21818268@zju.edu.cn (J.W.); niesheng@163.com (S.N.); 22118158@zju.edu.cn (X.D.); nbyysj@sina.com (J.S.)

[2] Key Laboratory of Precision Medicine for Atherosclerotic Diseases of Zhejiang Province, Ningbo 315010, China

[3] Medical Research Center, Ningbo First Hospital, Ningbo 315010, China

* Correspondence: huangy102@gmail.com (Y.H.); qinyuecui@163.com (X.G.)

Abstract: Background: Glioblastoma (GBM) is the most common and deadly brain tumor. The clinical significance of necroptosis (NCPS) genes in GBM is unclear. The goal of this study is to reveal the potential prognostic NCPS genes associated with GBM, elucidate their functions, and establish an effective prognostic model for GBM patients. Methods: Firstly, the NCPS genes in GBM were identified by single-cell analysis of the GSE182109 dataset in the GEO database and weighted co-expression network analysis (WGCNA) of The Cancer Genome Atlas (TCGA) data. Three machine learning algorithms (Lasso, SVM-RFE, Boruta) combined with COX regression were used to build prognostic models. The subsequent analysis included survival, immune microenvironments, and mutations. Finally, the clinical significance of NCPS in GBM was explored by constructing nomograms. Results: We constructed a GBM prognostic model composed of NCPS-related genes, including CTSD, AP1S1, YWHAG, and IER3, which were validated to have good performance. According to the above prognostic model, GBM patients in the TCGA and CGGA groups could be divided into two groups according to NCPS, with significant differences in survival analysis between the two groups and a markedly worse prognostic status in the high NCPS group ($p < 0.001$). In addition, the high NCPS group had higher levels of immune checkpoint-related gene expression, suggesting that they may be more likely to benefit from immunotherapy. Conclusions: Four genes (CTSD, AP1S1, YWHAG, and IER3) were screened through three machine learning algorithms to construct a prognostic model for GBM. These key and novel diagnostic markers may become new targets for diagnosing and treating patients with GBM.

Keywords: glioblastoma; necroptosis; machine learning algorithms; single-cell analysis; immune microenvironments

Citation: Wu, Y.; Huang, Y.; Zhou, C.; Wang, H.; Wang, Z.; Wu, J.; Nie, S.; Deng, X.; Sun, J.; Gao, X. A Novel Necroptosis-Related Prognostic Signature of Glioblastoma Based on Transcriptomics Analysis and Single Cell Sequencing Analysis. *Brain Sci.* **2022**, *12*, 988. https://doi.org/10.3390/brainsci12080988

Academic Editor: Hailiang Tang

Received: 25 June 2022
Accepted: 25 July 2022
Published: 26 July 2022

Publisher's Note: MDPI stays neutral with regard to jurisdictional claims in published maps and institutional affiliations.

1. Introduction

Glioblastoma (GBM) is an aggressive, highly malignant, and lethal brain tumor that originates from astrocytes and accounts for approximately 80% of all CNS malignancies [1]. The overall survival of GBM patients after standard treatment is generally less than 15 months, and although some patients with recurrent GBM can benefit from improved survival and prognosis when they undergo secondary surgery [2,3], the overall 5-year survival rate is still less than 5% [4,5]. The high frequency of molecular mutations (p53, RB, unmethylated MGMT) makes GBM less sensitive to conventional cytotoxic therapies and highly resistant to standard chemotherapy and radiotherapy [6]. Thus, it is important to identify new targets and molecular mechanisms that can propose alternative therapeutic strategies. Genetic data mining has attracted more and more attention with the development of high-throughput sequencing and bioinformatics. Therefore, the discovery of new markers that inhibit cell growth, migration, and invasion in GBM, and the study of their biological mechanisms for treating GBM patients, as well as the construction of prognostic models are undoubtedly essential for the advancement of prognosis and treatment.

Tumors can escape from immune surveillance, which includes the loss of programmed apoptosis [7]. Therefore, the inducing of programmed apoptosis in tumor cells is an important treatment for tumors. Targeting NCPS is considered a prospective cancer treatment strategy. Programmed cell death participated in homeostatic regulation, inflammatory response, anti-infective, and carcinogenic, which includes apoptosis, cell scorch, NCPS, and iron death [8–10]. Since the discovery of NCPS by HITOMI et al. in 2008 [11], the underlying mechanisms have been elucidated as research progresses [12]. Further exploration of NCPS has contributed to a new understanding of the composition of programmed cell death. It is mechanistically similar to apoptosis and morphologically similar to necrosis. It is mediated by activated receptor-interacting protein kinase 1 (RIPK1), RIPK3, and mixed-spectrum kinase domain-like (MLKL) [13]. Previous studies find NCPS associated with several tumor types and their immune microenvironment [14]. At present, research on NCPS in neurological disorders has focused on non-neoplastic diseases, such as cerebral ischemia [15], traumatic brain injury [16], and Alzheimer's disease [17]. Previous researchers have shown that Alzheimer's disease can be alleviated by overexpression of the core molecule of NCPS, MLKL [17]. Damage caused by cerebral ischemia can be reduced by inhibitors of NCPS [15]. However, its relationship with GBM is not yet clear. Thus, it is of great significance to explore the role of NCPS in GBM.

Single-cell RNA sequencing (scRNA-seq) is now emerging as a new approach to the study of human tumors, allowing gene expression data to be analyzed at the individual cell level. The multidimensional analysis of cell clustering allows for a clear exploration of the cellular heterogeneity of tumor tissue and the spatial, molecular, and functional heterogeneity of associated immune cells [18]. Currently, some researchers have begun to explore glioma heterogeneity from a single-cell perspective [19–21]. However, the majority of studies have focused on the heterogeneity of immune cells themselves in gliomas, with limited studies focusing on the expression of cell death-related molecules in gliomas and the impact on the immune microenvironment, particularly NCPS.

Nowadays, next-generation sequencing (NGS) technologies allow us to explore the variation of thousands of molecules at multiple omics levels, but how to choose the right and accurate variables in a high-dimensional computation, machine learning is a promising algorithm for computing such problems [22]. Some of the popular machine learning algorithms include the Least Absolute Shrinkage and Selection Operator (LASSO), Support Vector Machine (SVM), and Random Forest (RF). LASSO [23] is a regression method that is of great importance in datasets with many variables and supports unbiased parameter selection. SVM is an algorithm that efficiently handles problems by using a non-linear kernel function [24,25], while recursive feature elimination (RFE) is a popular method of variable selection at the multi-omics level [26,27]. In this research, we used support vector machine recursive feature elimination (SVM-RFE) to dimensionally reduce the search for suitable variables. Random Forest is an integrated algorithm based on decision trees [28]. The Boruta algorithm is a wrapper around the Random Forest algorithm, allowing the algorithm to iteratively remove less relevant features determined by statistical tests [29].

In this study, to explore GBM-associated potentially prognostic NCPS genes and clarify their functions to provide a new reference for the diagnosis and treatment of GBM patients, we collected and collated GBM patients' data from The Cancer Genome Atlas (TCGA) database, the GEO database, and sequencing data from China Glioma Genome Atlas Database (CGGA). Weighted co-expression network analysis (WGCNA) combined with machine learning was used to develop a GBM NCPS-related prognostic model, which was divided into two groups based on median values according to risk scores. In addition, the value of using multiple immune scoring algorithms to assess the tumor immune microenvironment and tumor mutational load characteristics is also explored in this paper.

2. Materials and Methods

2.1. Transcriptome Data Download and Processing

The transcriptome files of GBM and the clinical information of the samples corresponding to them were downloaded from the TCGA database (https://portal.gdc.cancer.gov/, accessed on 16 May 2022), and the TPM data were extracted for subsequent analysis. After removing the non-primary tumor samples, a final sample of 143 GBM patients with complete clinical information was used as a training cohort (Table S1). After downloading 325 glioma samples from the CGGA portal (http://www.cgga.org.cn/index.jsp, accessed on 16 May 2022), only GBM-related information was retained, and, finally, 85 GBM samples were obtained, and the transcriptome FPKM data were converted to TPM data and combined with clinical information as the validation cohort data [30].

2.2. Single-Cell Data Download and Processing

The GSE182109 single-cell glioma dataset was downloaded from the GEO database (https://www.ncbi.nlm.nih.gov/geo/, accessed on 19 May 2022) and 11 GBM samples from 44 glioma patients were selected for analysis [31]. To assure data quality, the cells selected for this study set the screening criteria as less than 20% of mitochondrial genes, more than 200 genes, and gene expression between 200 and 7000, and containing more than three genes. The number of highly variant genes was 3000. Samples were integrated using SCT correction. Next, data dimensionality was reduced by setting the "DIMS" parameter to 25 and using the TSNE method. The clustering of cells was performed using the "KNN" method. Cells were subsequently annotated using different cell surface markers. Finally, NCPS genes were imported using the "AddModuleScore" function to determine the percentage of NCPS-related genes in each cell.

2.3. The Acquisition of NCPS-Related Genes

A total of 614 genes associated with NCPS were identified in the Genecards database (https://www.genecards.org/, accessed on 17 April 2022). We set the correlation coefficient threshold to greater than or equal to 1.0 and finally obtained 91 genes associated with NCPS.

2.4. Weighted Co-Expression Network Analysis (WGCNA) and Single Sample Gene Set Enrichment Analysis (ssGSEA)

Weighted gene co-expression network analysis (WGCNA) is a systems biology approach that explores the gene network itself and its associated phenotypes, as well as the core genes in the network, by finding highly correlated gene modules, which are ultimately used to identify candidate biomarker genes or therapeutic targets. Additionally, ssGSEA analysis is commonly used to quantify specific genomes in a sample. In this study, ssGSEA analysis was used to obtain the associated scores of NCPSfor each GBM patient. Then, WGCNA was used to obtain GBM score-associated gene modules in order to identify NCPS-associated genes.

2.5. Construction of the Prognostic Model Associated with NCPS

Univariate Cox analysis was used to initially identify the correlation between NCPS-related genes and patient prognosis, and we selected a *p* value less than 0.01 as an indication that the gene had a significant prognostic ability. After filtering, candidate prognostic genes were selected by a combined analysis of three machine learning algorithms, including the LASSO regression algorithm, the SVM-RFE algorithm [32], and the random forest-based Boruta algorithm [33]. In all three algorithms, the number of random seeds is set to 2022, the maximum number of passes of the lambda value on the data in LASSO is 1000, the 10-fold cross-validation is used in the RFE algorithm and the calculation method is set to "svmRadial", and the number of trees to 500 is adjusted in the Boruta algorithm. Subsequently, overlapping genes from the three algorithms were selected from the candidate genes and, finally, a prognostic model was constructed using multivariate Cox regression. The correlation coefficients from multifactorial Cox regression analyses allowed us to derive the NCPS scores for each sample and to divide them into two groups according to the median. The prognosis was assessed using the Kaplan–Meier curve, and $p < 0.05$ was considered a clinically significant prognosis. The classification of GBM was evaluated by applying principal component analysis (PCA) to observe the grouping of this model. On this basis, modeling tests were performed on the external data using the above-mentioned methods.

2.6. Immunological Function and Mutation Analysis

The "IBOR" package integrates various immune infiltration assessment algorithms in the R language [34]. Using the "IBOR" package, we evaluated seven immune infiltration algorithms for GBM patients in the TCGA repository and showed the different levels of infiltration of various immune cells in a heat map to investigate the differences in immune cell infiltration between the different necroptotic groups. The genes associated with the immune checkpoint identified in the NCPS subgroup were also analyzed in a box plot. We also performed a study of the major variants in GBM, presenting the top 20 genes with the highest mutation frequency and the base types of the major mutations in GBM.

2.7. Location and Expression of Prognostic Models in Single-Cell Sequencing Analysis

In single-cell sequencing analysis, the expression of the genes used to construct the prognostic model was analyzed in the well-annotated cells to find out the expression of a specific gene in prognostic model genes at the single-cell level, and the expression was presented using TSNE plots.

2.8. Building a Predictive Nomogram

In this study, we used the Nomogram to establish a method that can effectively predict the probability of survival of patients, taking into account the influence of clinical covariates. The NCPS values were correlated with clinical data to construct a Nomogram to evaluate the risk of death in patients with GBM. Finally, the patient's prognosis was evaluated based on the ROC curve.

3. Results

3.1. Single Cell Sequencing Data Analysis

The study flow chart of this article is shown in Figure 1. The single-cell sequencing dataset GSE182109 from GBM was used as a follow-up analysis and different samples were integrated. The results are shown in Figure 2A, with 11 samples having no significant batch effects for subsequent analysis. All cells were then clustered into 12 clusters using the KNN clustering algorithm (Figure 2B,C). Subsequently, using the "AddModuleScore" function, 91 genes associated with NCPS were entered, resulting in the percentage of each cell type associated with NCPS. Based on the percentage of cells obtained, these cells were divided into two groups according to the median value and displayed as TSNE plots (Figure 2D). Then, based on the surface marker genes of different cell types (Table S2), their expression in different clusters was observed (Figure 2C), and, finally, seven cell types were identified based on the expression, they were: B cells, T cells, endothelial cells, myeloid cells, pericytes, oligodendrocytes, and glioblastoma cells (Figure 2E). The differentially expressed genes were then analyzed for the high and low NCPS groups, and a total of 2451 differential genes were identified.

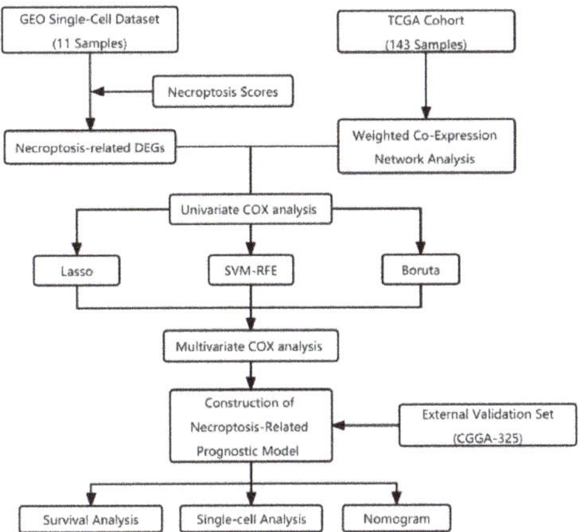

Figure 1. Flow chart.

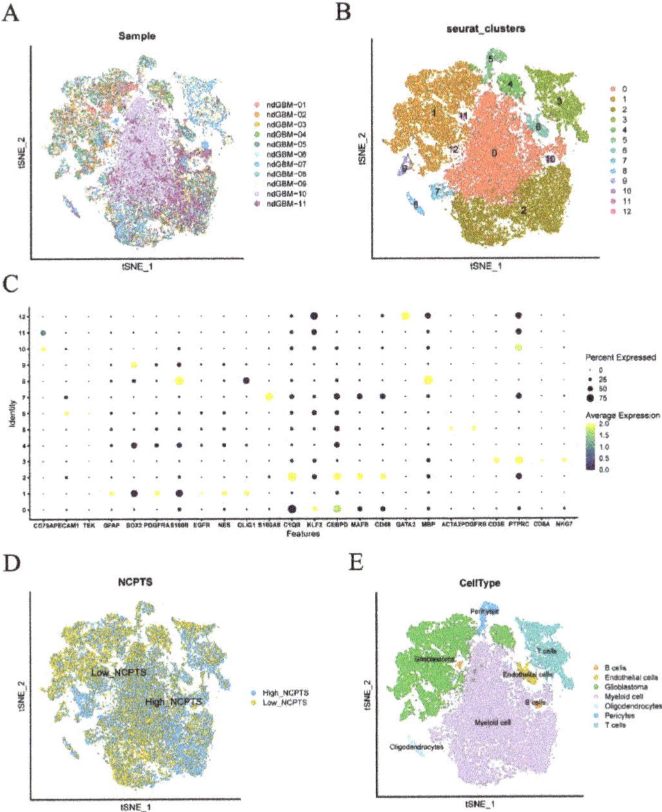

Figure 2. Single-cell sequencing analysis of GSE182109. (**A**) The integration effect of 11 samples is good. (**B,C**) Dimensionality reduction and cluster analysis. All cells in 11 samples were clustered into 12 clusters. (**D**) The percentage of necroptosis genes in each cell. The cells were divided into high- and low-necroptosis cells. (**E**) According to the surface marker genes of different cell types, the cells are annotated as B cells, T cells, endothelial cells, myeloid cells, pericytes, oligodendrocytes, and glioblastoma cells, respectively.

3.2. Weighted Co-Expression Network Analysis

TCGA transcriptomics were performed in 143 GBM patients using the WGCNA method to acquire gene modules related to NCPS. The soft threshold was set to 12, the minimum number of modules was set to 80, deepSplit was set to 2, and the modules with similarity lower than 0.25 were merged to derive a total of 15 non-grey modules (Figure 3A). In non-grey modules, a strong association was found between Meblue, Mebrown, and Meyellow and NCPS (Figure 3B). Genes from the three modules were selected for subsequent analysis.

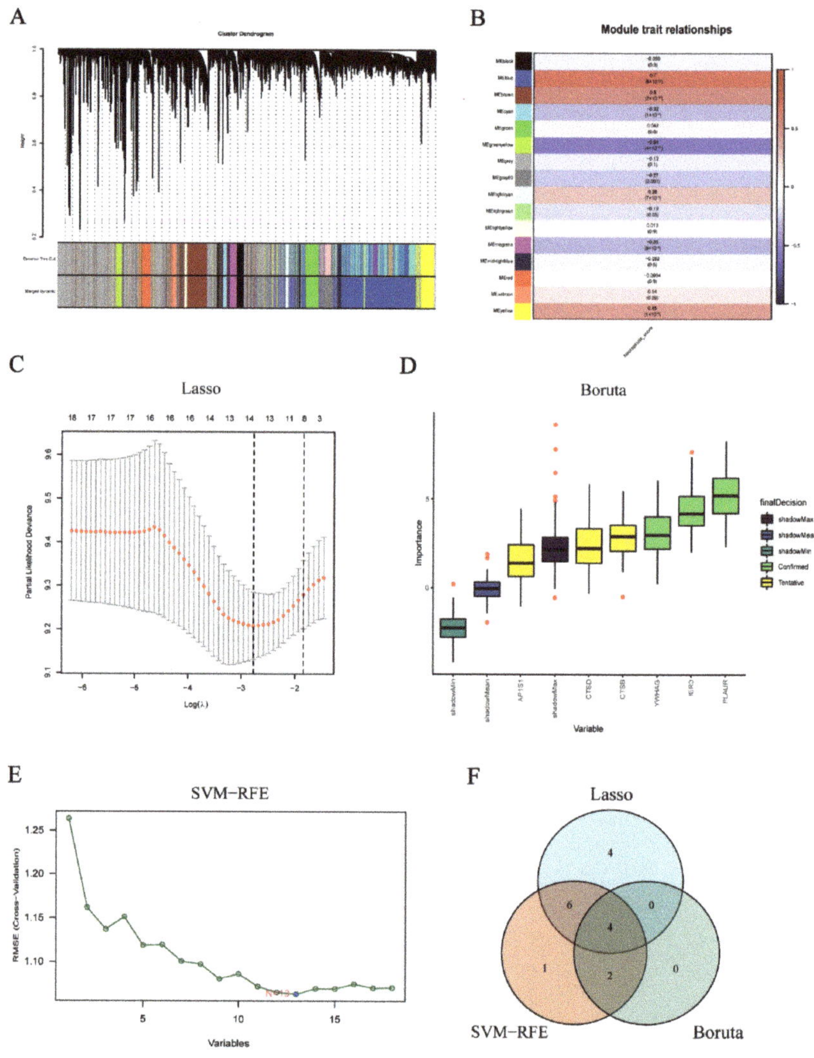

Figure 3. WGCNA and construction of the necroptosis-related prognostic model. (**A**,**B**) WGCNA found that Meblue, Mebrown, and Meyellow modules were closely related to the score of necroptosis. (**C**) Fourteen genes were selected to construct the prognostic model by Lasso regression. (**D**) Six genes were selected to construct the prognostic model by the Boruta algorithm. (**E**) Thirteen genes were selected to construct the prognostic model by the SVM-RFE algorithm. (**F**) Intersection feature selection between LASSO, SVM-RFE algorithm, and Boruta algorithm and their individual components.

3.3. Selection of Prognostic Candidate Genes by Machine Learning Algorithms

To begin with, differential genes between high and low expression populations were acquired according to the NCPS gene scores from the previous single-cell dataset. It was then intersected with the NCPS-related genes obtained in the WGCNA to yield a total of 1437 genes. By matching the TCGA with the crossover genes obtained in the previous step, eventually, all genes analyzed in the previous step were allowed for subsequent calculations. In the following analysis, we take the TCGA cohort as our training set, univariate Cox analysis was set at $p < 0.01$, and 18 genes associated with patient prognosis were initially obtained. Then, the random seed was set to 2022 and the results of the regression revealed that the contraction of the lasso stabilized when the number of variables was 14, with the smallest deviation in the score and the best LAMDA of 0.062 (Figure 3C). The Boruta algorithm set the same random seeds and outcome variables as the previous two algorithms, and obtained three genes that the algorithm considered as definitively correlated with survival time, called "Confirmed"; three genes that were considered potentially correlated, described as "Tentative"; the rest of the variables were not considered relevant. In order to include more variables, "Confirmed" and "Tentative" were selected for inclusion in the final results and plotted for display (Figure 3D). SVM-RFE used survival time as the outcome variable, and the results of the 10-fold validation showed that minimal root mean square error (RMSE) was obtained when the number of genes was 13, outputting the above genes as the result of SVM-RFE (Figure 3E). The results of the above algorithms were presented in Venn diagrams and the overlapping results were selected as the final candidate features for inclusion in the construction of the model so that we obtained four genes (Figure 3F). They are CTSD, AP1S1, YWHAG, and IER3, respectively. The respective optimal results of these three algorithms and the intersecting genes are summarized in Table S3.

3.4. Construction and Validation of NCPS-Related Prognostic Model

Based on the results of multivariate Cox regression analysis, the four genes previously analyzed were still chosen to construct prognostic signatures (Figure 4A). For this signature, the concordance index = 0.65, Akaike information criterion (AIC) = 874.69, and p-value = 4.55×10^{-6}. The following equation was applied to construct the risk score formula: NCPS = (0.002758577 * Expr CTSD) + (0.00431492 * Expr AP1S1) + (0.001704931 * Expr YWHAG) + (0.013004668 * Expr IER3). Where each coefficient is derived from the results of multivariate Cox regression. All four genes in the formula were considered risk factors for HR > 1 (Figure 4A). A score was assigned to the prognosis of each GBM patient based on the above formula, and all patients were divided into two groups, named the high-risk or low-risk group, according to the median of this score. In the figure, survival time was more beneficial for patients with low-risk scores ($p < 0.001$). Surprisingly, in the validation set (CGGA), we observed similar results, namely a significantly worse outcome in patients with high NCPS ($p < 0.01$, Figure 4B,C). Furthermore, we analyzed the ROC curves of the training and validation cohorts (Figure S1). Finally, PCA analysis was used to analyze the high-risk and low-risk of the training and validation groups, which indicated that the model could better group the GBM patients. (Figure 4D,E).

3.6. Cell Localization of Four Modeling Genes

After constructing the model, we tried to obtain more insight into the expression of the screened NCPS-related genes in different cell types, which we explored in single-cell sequencing analysis. As shown in Figure 6A–F, CTSD expression was highest in myeloid cells, high expression of AP1S1 was shown in glioblastoma cells, and YWHAG was mainly expressed in glioblastoma cells, oligodendrocytes, and myeloid cells, and IER3 was predominantly expressed in myeloid cells.

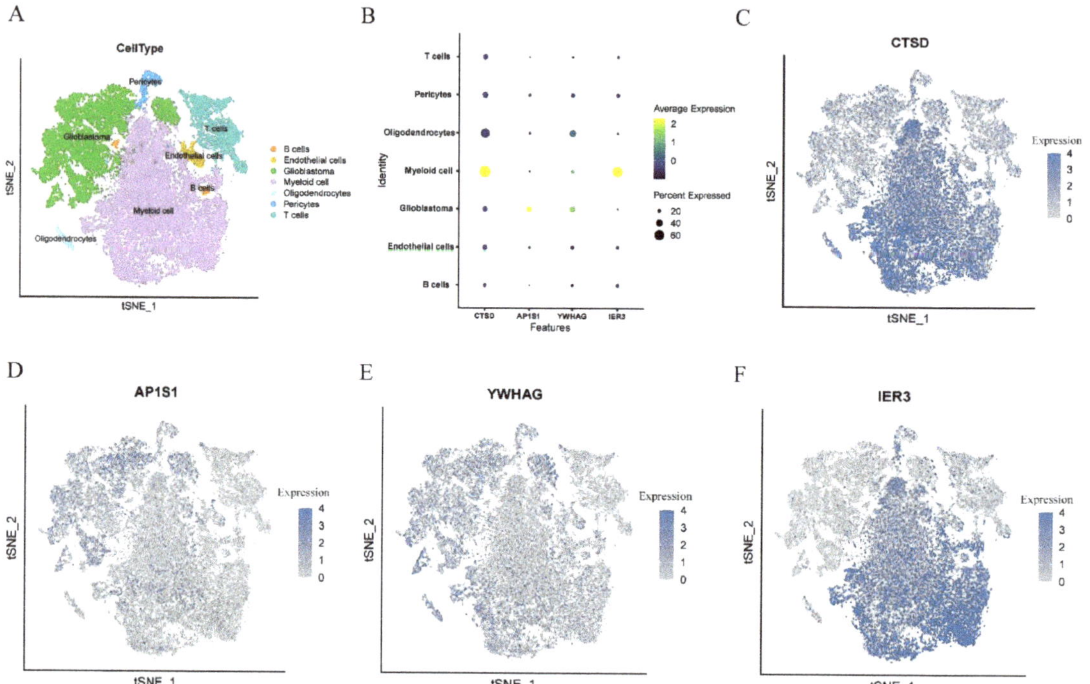

Figure 6. Single-cell sequencing analysis to explore the cell localization of 4 modeling genes. (**A**) The location of different cell types in GBM. (**B–F**) CTSD expression was highest in myeloid cells, high expression of AP1S1 was shown in glioblastoma cells, YWHAG was mainly expressed in glioblastoma cells, oligodendrocytes, and myeloid cells, and IER3 was predominantly expressed in myeloid cells.

3.7. The Construction of a Nomogram

By analyzing the clinical data and NCPS values, a nomogram was constructed to better evaluate the risk of GBM patients (Figure 7A). The mortality rates at 1, 2, and 3 years were estimated from the "TCGA-02-0047" patients by gender, age, race, and NCPS scores of 0.369, 0.781, and 0.931, respectively. The nomogram can better evaluate the patient's risk and guide future clinical decisions. This method was analyzed using the ROC method to evaluate its model accuracy (Figure 7B). The AUC values were found to be 0.74, 0.76, and 0.78 for 1, 2, and 3 years. In addition, a decision curve analysis was also performed based on the area of 1, 2, and 3 years and the horizontal axis of None. The results indicate that this method is effective in predicting the prognostic survival time of patients and has some reference value for clinical treatment decisions (Figure 7C).

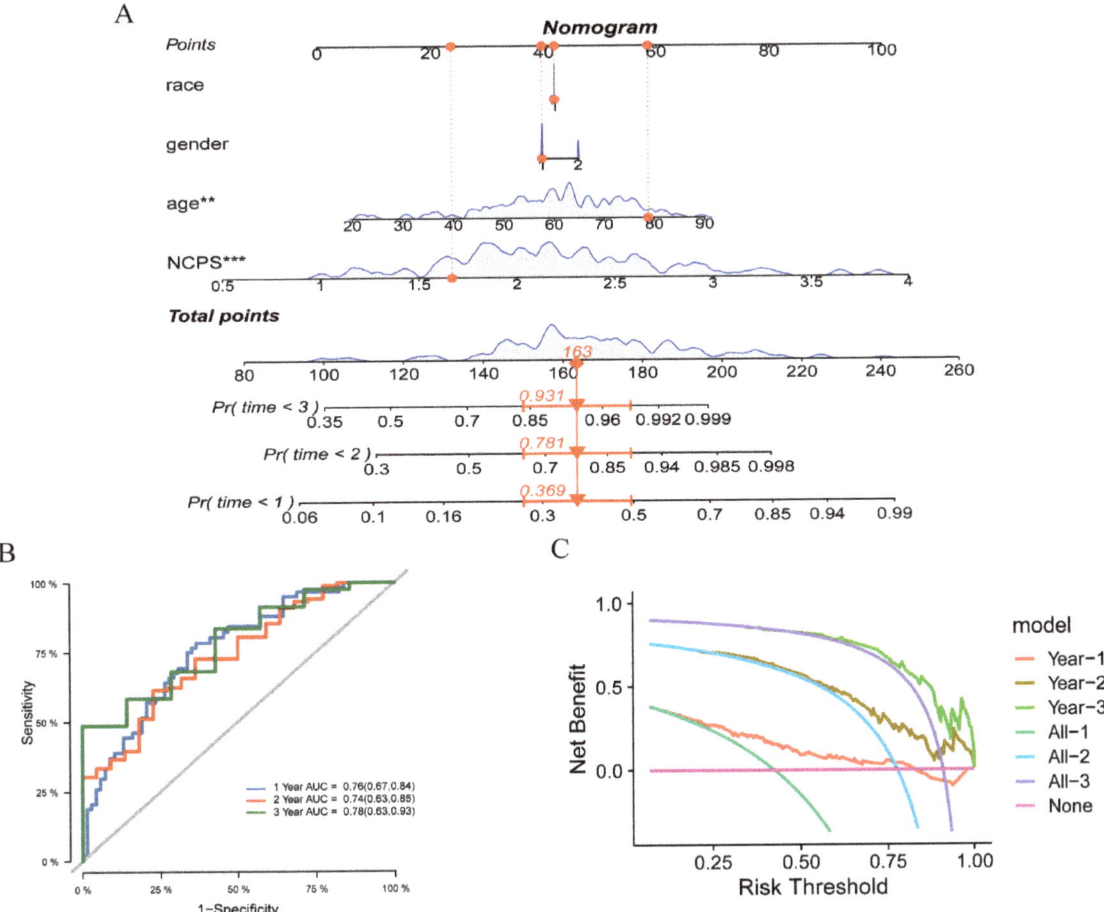

Figure 7. The construction of a nomogram. (**A**) Nomogram of patient "TCGA-02-0047". The mortality rates at 1, 2, and 3 years were estimated from the "TCGA-02-0047" patients by gender, age, race, and NCPS scores of 0.369, 0.781, and 0.931, respectively. (**B**) ROC curve of the nomogram. The area under the curve (AUC) in 1, 2, and 3 years were 0.76, 0.74, and 0.78, respectively. (**C**) Decision curve analysis. This method is effective in predicting the prognostic survival time of patients. ** $p < 0.01$; *** $p < 0.001$.

4. Discussion

Despite partial clinical advances, GBM remains the most aggressive and common malignancy of the brain [35]. Due to heterogeneity, rapid appreciation, and aggressiveness, patients often have a low average survival of 12–15 months [36]. An increasing number of patients with advanced solid tumors have seen some improvement in prognosis, stemming from the translation of immunotherapy to clinical oncology. GBM is unlike other extracranial solid tumors, as peripheral conventional immune cells seem to be less appreciated intracranially and the brain is mostly in a state of immune quiescence; on the other hand, the brain becomes very poorly tolerated when an intracranial lesion arises, especially when an immune attack from a tumor-like GBM occurs, making it fascinating to explore the immune microenvironment of GBM [37]. In the current study, we constructed a prognostic signature associated with NCPS based on machine learning algorithms through extensive analysis of GBM in single-cell sequencing data and TCGA transcriptome data. Four genes were finally screened as having prognostic values, namely CTSD, AP1S1, YWHAG, and

IER3. We then performed a risk score on their constructed models to classify GBM patients into high- and low-risk groups. The results of the survival curves explicitly showed that patients with NCPS-related high risk scores had a worse prognosis ($p < 0.001$).

NCPS plays a critical role in the regulation of cancer biology, including tumorigenesis and cancer metastasis [38]. Therapeutic strategies targeting apoptosis resistance in tumor cells have led to the increasing research status of NCPS in tumors. The specific process of NCPS involves activation of RIP1 by TNF-α/TNFR1 when caspase-8 is cleared or inhibited by mutation or infection and the subsequent recruitment of RIP3 through the RIP homotypic interaction motif (RHIM) [39]. The interaction of these two proteins results in the phosphorylation of RIP3, which activates its kinase in proper sequence, resulting in increased phosphorylation of RIP3 and downstream MLKL. The combinatorial interaction of RIP1–RIP3 and RIP3–MLKL collectively forms a necrosome complex to execute NCPS [40,41]. Several studies have shown that the effect of NCPS on tumors appears to be a double-edged sword: on the one hand, key molecules associated with NCPS promote cancer progression and reduce patient survival [42,43], and on the other hand, several studies have demonstrated that promoting NCPS can effectively inhibit tumor growth [44–46]. Although studies on the effects of NCPS on CNS tumor processes are limited, the available information suggests a possible link between genes related to the necroptotic pathway and the clinical behavior of GBM. For example, a significant reduction in overall survival (OS) in GBM under P53-induced inhibition is inextricably linked to the high expression of RIPK1 [47]. Nevertheless, the effect of NCPS on GBM has not been fully elucidated. To date, studies related to NCPS in GBM have been relatively scarce. In the current study, our results showed that four NCPS-related genes have predictable value in the prognosis of the GBM risk. The survival curves explicitly showed that patients with NCPS-related high risk scores had a worse prognosis. This suggests that NCPS may play a pro-cancer role in GBM, which is consistent with the results of other analyses [47,48]. Our study provided a basis for the prediction of NCPS-related genes in GBM and lays a foundation for further study of their prognosis.

The tumor microenvironment plays a pivotal role in tumors. In addition to the process of tumor development, it also has a remarkable impact on immunotherapy outcomes and overall patient survival [49]. In addition to the influence of the tumor and its surrounding microenvironment, the systemic immune-inflammatory status has also received increased attention [50,51]. Studies have found that elevated neutrophils, lymphocytes, and platelets in the blood of patients with tumors are associated with aggressiveness and poor prognosis of solid tumors [52,53]. In studies related to GBM, systemic immune-inflammatory index, which uses blood cells as a marker for detection, has been proven to be significantly associated with patient prognosis [54]. Interestingly, this index can also assess the impact of different types of surgery on the prognosis of GBM patients [55]. The complicated immune environment, involving immune cell distribution within and around the tumor, immune cell composition, and the overall immune environment of the GBM, can all have an impact on the effectiveness of immunotherapy and even the malignancy of the tumor [56]. We have researched the immune microenvironment of GBM and found that NCPS may be associated with macrophage infiltration in GBM, particularly in the M1 and M2 subtypes. Previous studies have also demonstrated the close association of macrophages not only with glioma survival and prognosis but also with their role in apoptosis [57]. Furthermore, in our follow-up study, we found that patients with highly expressed genes related to immune detection sites are more sensitive to this immunotherapy. This indicates that the model we made not only could determine the survival prognosis of patients but also could reflect the immune infiltration of the tumor. In addition, the comparison results between different risk groups showed that the number of mutated genes in the high-risk group was significantly higher than that in the low-risk group. This may be due to the deterioration of the microenvironment due to the mutation of immune-related genes, which, in turn, promotes tumorigenesis. Therefore, our results may provide new evidence for the immunotherapy of GBM.

The CTSD gene, a member of the peptidase A1 family, has been shown to be highly expressed in the breast cancer tumor microenvironment and is strongly associated with the survival of breast cancer patients [58]. Recent experiments have identified a role for CTSD in GBM: upregulation of CTSD expression contributes to enhanced radio-resistance of GBM cells, while inhibition of CTSD significantly reduces the migration ability of GBM cells [59]. As a member of the bridging protein (AP) family that coordinates various transports in the intracellular membrane pathway, AP1S1 and its related genes were shown to be associated with immune response, lysosomal vesicle transport, and protein presentation [60]. In previous bioinformatics analyses, investigators found that AP1S1 was highly expressed in GBM and associated with survival prognosis [61]. Our findings, which identified AP1S1 as an independent prognostic indicator in GBM, are consistent with previous studies. YWHAG, one of the 14-3-3 family isoforms, is involved as a regulatory molecule in a variety of cellular processes, including cell survival and apoptosis, protein transport, and cell cycle regulation [62,63]. Researchers have found that YWHAG was significantly overexpressed in GBM, head, and neck squamous cell carcinoma, and breast and gastric cancer [64–66]. IER3 belongs to a stress-inducible gene, also known as IEX-1, which is classified as a family of immediate early response genes. IER3 plays a complex, and to some extent contradictory, role in cell cycle control and apoptosis [67]. Current studies on IER3 in tumors have also confirmed that it is expressed in opposite ways in different tumors [68,69], so the exact mechanism and function of this gene remains to be explored. In this study, the results showed that CTSD and IER3 were predominantly expressed in myeloid cells, high expression of AP1S1 was shown in glioblastoma cells, and YWHAG was mainly expressed in glioblastoma cells, oligodendrocytes, and myeloid cells. This corroborates our previous evidence of association of NCPS with macrophages in immune infiltration. The value shown by these four genes in GBM patients deserves deeper investigation.

The previously published GSE182109 dataset explored the heterogeneity and immune infiltration of gliomas, including low-grade gliomas, and primary and recurrent glioblastomas, by single-cell sequencing analysis. The investigators analyzed the GBM immune microenvironment and classification of functional immune cell subtypes, on the basis of which specific immune checkpoints were inspected and screened, and S100A4 was identified as a potential target for immunotherapy [31]. In this study, primary glioblastoma data from the GSE182109 dataset were first subjected to further single-cell analysis, which divided them into two groups with different NCPS states. This is a useful reference for studying the heterogeneity of NCPS states in GBM. Afterward, a prognostic model was constructed based on differentially expressed genes in the cell population. This model was tested using TCGA survival data. One of the strengths of this model is its ability to accurately assess the prognosis of GBM patients. This study is the first to apply single-cell clustering analysis to predict the prognosis of GBM cell necroptosis, which may not only provide a new perspective to study its regulated cell death different from transcriptome analysis, but also provide a new angle for clinical treatment. Admittedly, there are certain shortcomings in the present study. Due to the limitations of current technology, GBM patients are still under-explored at the single-cell level on various immune cell subtypes with NCPS. In addition, there is no standardized assay for clinical screening of the expression of the NCPS-related genes we obtained, and we know too little about the underlying mechanisms of the complex functions of NCPS, which we hope will be investigated more in the future scientific work.

5. Conclusions

In summary, we developed a predictive model for glioblastoma NCPS genes. These genes may be closely related to the mechanism of NCPS in GBM. The expression of these four genes in specific cells, such as myeloid cells and glioblastoma cells, makes us more curious about the specific function of NCPS on immune cells and tumor cells. These results contribute to further understanding of the molecular mechanisms of GBM regulation. The construction of this model allows us to evaluate GBM from a new perspective, in addition to the tumor microenvironment of GBM, and to design effective combination therapies based on cell type-specific expression.

Supplementary Materials: The following supporting information can be downloaded at: https://www.mdpi.com/article/10.3390/brainsci12080988/s1, Figure S1: ROC curves of the training and validation cohorts; Table S1: Baseline of GBM patients; Table S2: Cell Types Annotation; Table S3: Machine learning algorithms and results.

Author Contributions: Conceptualization, X.G. and J.S.; methodology, Y.W., Y.H. and J.W.; software, Y.W.; validation, C.Z. and H.W.; formal analysis, Z.W.; investigation, Z.W.; writing—original draft preparation, Y.W.; writing—review and editing, Y.H.; supervision, S.N. and X.D.; project administration, J.S.; funding acquisition, X.G. All authors have read and agreed to the published version of the manuscript.

Funding: This research was funded by the grants from the Zhejiang Provincial Natural Science Foundation of China (LY22H090001), Medicine and health science and technology projects of Zhejiang province (2022KY305, 2022KY322), Ningbo Health Branding Subject Fund (PPXK2018-04), Ningbo Science and Technology Innovation 2025 Major Project (2022Z125), Key Laboratory of Precision Medicine for Atherosclerotic Diseases of Zhejiang Province (2022E10026).

Institutional Review Board Statement: Not applicable.

Informed Consent Statement: Not applicable.

Data Availability Statement: Data are contained within the article.

Conflicts of Interest: The authors declare no conflict of interest.

References

1. Pagano, C.; Navarra, G.; Coppola, L.; Avilia, G.; Pastorino, O.; Della Monica, R.; Buonaiuto, M.; Torelli, G.; Caiazzo, P.; Bifulco, M.; et al. N6-isopentenyladenosine induces cell death through necroptosis in human glioblastoma cells. *Cell Death Discov.* **2022**, *8*, 173. [CrossRef] [PubMed]
2. Pasqualetti, F.; Montemurro, N.; Desideri, I.; Loi, M.; Giannini, N.; Gadducci, G.; Malfatti, G.; Cantarella, M.; Gonnelli, A.; Montrone, S.; et al. Impact of recurrence pattern in patients undergoing a second surgery for recurrent glioblastoma. *Acta Neurol. Belg.* **2022**, *122*, 441–446. [CrossRef] [PubMed]
3. Suchorska, B.; Weller, M.; Tabatabai, G.; Senft, C.; Hau, P.; Sabel, M.C.; Herrlinger, U.; Ketter, R.; Schlegel, U.; Marosi, C.; et al. Complete resection of contrast-enhancing tumor volume is associated with improved survival in recurrent glioblastoma-results from the DIRECTOR trial. *Neuro Oncol.* **2016**, *18*, 549–556. [CrossRef] [PubMed]
4. Lah, T.T.; Novak, M.; Breznik, B. Brain malignancies: Glioblastoma and brain metastases. *Semin. Cancer Biol.* **2020**, *60*, 262–273. [CrossRef]
5. Stupp, R.; Brada, M.; van den Bent, M.J.; Tonn, J.C.; Pentheroudakis, G. High-grade glioma: ESMO Clinical Practice Guidelines for diagnosis, treatment and follow-up. *Ann. Oncol.* **2014**, *25* (Suppl. S3), iii93–iii101. [CrossRef]
6. Akyerli, C.B.; Yüksel, Ş.; Can, Ö.; Erson-Omay, E.Z.; Oktay, Y.; Coşgun, E.; Ülgen, E.; Erdemgil, Y.; Sav, A.; von Deimling, A.; et al. Use of telomerase promoter mutations to mark specific molecular subsets with reciprocal clinical behavior in IDH mutant and IDH wild-type diffuse gliomas. *J. Neurosurg.* **2018**, *128*, 1102–1114. [CrossRef]
7. Swann, J.B.; Smyth, M.J. Immune surveillance of tumors. *J. Clin. Investig.* **2007**, *117*, 1137–1146. [CrossRef]
8. Peñaranda Fajardo, N.M.; Meijer, C.; Kruyt, F.A.E. The endoplasmic reticulum stress/unfolded protein response in gliomagenesis, tumor progression and as a therapeutic target in glioblastoma. *Biochem. Pharmacol.* **2016**, *118*, 1–8. [CrossRef]
9. Zheng, P.; Ding, B.; Zhu, G.; Li, C.; Lin, J. Biodegradable Ca^{2+} Nanomodulators Activate Pyroptosis through Mitochondrial Ca^{2+} Overload for Cancer Immunotherapy. *Angew. Chem. Int. Ed. Engl.* **2022**. [CrossRef]
10. de Vasconcelos, N.M.; Van Opdenbosch, N.; Van Gorp, H.; Parthoens, E.; Lamkanfi, M. Single-cell analysis of pyroptosis dynamics reveals conserved GSDMD-mediated subcellular events that precede plasma membrane rupture. *Cell Death Differ.* **2019**, *26*, 146–161. [CrossRef]
11. Galluzzi, L.; Kroemer, G. Necroptosis: A specialized pathway of programmed necrosis. *Cell* **2008**, *135*, 1161–1163. [CrossRef] [PubMed]
12. Su, Z.; Yang, Z.; Xu, Y.; Chen, Y.; Yu, Q. Apoptosis, autophagy, necroptosis, and cancer metastasis. *Mol. Cancer* **2015**, *14*, 48. [CrossRef] [PubMed]
13. Krysko, O.; Aaes, T.L.; Kagan, V.E.; D'Herde, K.; Bachert, C.; Leybaert, L.; Vandenabeele, P.; Krysko, D.V. Necroptotic cell death in anti-cancer therapy. *Immunol. Rev.* **2017**, *280*, 207–219. [CrossRef] [PubMed]

14. Gong, Y.; Fan, Z.; Luo, G.; Yang, C.; Huang, Q.; Fan, K.; Cheng, H.; Jin, K.; Ni, Q.; Yu, X.; et al. The role of necroptosis in cancer biology and therapy. *Mol. Cancer* **2019**, *18*, 100. [CrossRef]
15. Chen, Y.; Zhang, L.; Yu, H.; Song, K.; Shi, J.; Chen, L.; Cheng, J. Necrostatin-1 Improves Long-term Functional Recovery Through Protecting Oligodendrocyte Precursor Cells After Transient Focal Cerebral Ischemia in Mice. *Neuroscience* **2018**, *371*, 229–241. [CrossRef]
16. Liu, Z.-M.; Chen, Q.-X.; Chen, Z.-B.; Tian, D.-F.; Li, M.-C.; Wang, J.-M.; Wang, L.; Liu, B.-H.; Zhang, S.-Q.; Li, F.; et al. RIP3 deficiency protects against traumatic brain injury (TBI) through suppressing oxidative stress, inflammation and apoptosis: Dependent on AMPK pathway. *Biochem. Biophys. Res. Commun.* **2018**, *499*, 112–119. [CrossRef] [PubMed]
17. Caccamo, A.; Branca, C.; Piras, I.S.; Ferreira, E.; Huentelman, M.J.; Liang, W.S.; Readhead, B.; Dudley, J.T.; Spangenberg, E.E.; Green, K.N.; et al. Necroptosis activation in Alzheimer's disease. *Nat. Neurosci.* **2017**, *20*, 1236–1246. [CrossRef]
18. Picelli, S.; Björklund, Å.K.; Faridani, O.R.; Sagasser, S.; Winberg, G.; Sandberg, R. Smart-seq2 for sensitive full-length transcriptome profiling in single cells. *Nat. Methods* **2013**, *10*, 1096–1098. [CrossRef]
19. Neftel, C.; Laffy, J.; Filbin, M.G.; Hara, T.; Shore, M.E.; Rahme, G.J.; Richman, A.R.; Silverbush, D.; Shaw, M.L.; Hebert, C.M.; et al. An Integrative Model of Cellular States, Plasticity, and Genetics for Glioblastoma. *Cell* **2019**, *178*, 835–849.e21. [CrossRef]
20. Lee, J.-K.; Wang, J.; Sa, J.K.; Ladewig, E.; Lee, H.-O.; Lee, I.-H.; Kang, H.J.; Rosenbloom, D.S.; Camara, P.G.; Liu, Z.; et al. Spatiotemporal genomic architecture informs precision oncology in glioblastoma. *Nat. Genet.* **2017**, *49*, 594–599. [CrossRef]
21. Darmanis, S.; Sloan, S.A.; Croote, D.; Mignardi, M.; Chernikova, S.; Samghababi, P.; Zhang, Y.; Neff, N.; Kowarsky, M.; Caneda, C.; et al. Single-Cell RNA-Seq Analysis of Infiltrating Neoplastic Cells at the Migrating Front of Human Glioblastoma. *Cell Rep.* **2017**, *21*, 1399–1410. [CrossRef] [PubMed]
22. Degenhardt, F.; Seifert, S.; Szymczak, S. Evaluation of variable selection methods for random forests and omics data sets. *Brief. Bioinform.* **2019**, *20*, 492–503. [CrossRef]
23. Tibshirani, R. Regression shrinkage and selection via the lasso. *J. R. Stat. Soc. Ser. B* **1996**, *58*, 267–288. [CrossRef]
24. Ng, K.L.S.; Mishra, S.K. De novo SVM classification of precursor microRNAs from genomic pseudo hairpins using global and intrinsic folding measures. *Bioinformatics* **2007**, *23*, 1321–1330. [CrossRef]
25. Rice, S.B.; Nenadic, G.; Stapley, B.J. Mining protein function from text using term-based support vector machines. *BMC Bioinform.* **2005**, *6* (Suppl. S1), S22. [CrossRef] [PubMed]
26. Habermann, J.K.; Doering, J.; Hautaniemi, S.; Roblick, U.J.; Bündgen, N.K.; Nicorici, D.; Kronenwett, U.; Rathnagiriswaran, S.; Mettu, R.K.R.; Ma, Y.; et al. The gene expression signature of genomic instability in breast cancer is an independent predictor of clinical outcome. *Int. J. Cancer* **2009**, *124*, 1552–1564. [CrossRef] [PubMed]
27. Dietrich, S.; Floegel, A.; Weikert, C.; Prehn, C.; Adamski, J.; Pischon, T.; Boeing, H.; Drogan, D. Identification of Serum Metabolites Associated With Incident Hypertension in the European Prospective Investigation into Cancer and Nutrition-Potsdam Study. *Hypertension* **2016**, *68*, 471–477. [CrossRef]
28. Breiman, L. Random forests. *Mach. Learn.* **2001**, *45*, 5–32. [CrossRef]
29. Kursa, M.B.; Rudnicki, W.R. Feature selection with the Boruta package. *J. Stat. Softw.* **2010**, *36*, 1–13. [CrossRef]
30. Zhao, Z.; Meng, F.; Wang, W.; Wang, Z.; Zhang, C.; Jiang, T. Comprehensive RNA-seq transcriptomic profiling in the malignant progression of gliomas. *Sci. Data* **2017**, *4*, 170024. [CrossRef]
31. Abdelfattah, N.; Kumar, P.; Wang, C.; Leu, J.-S.; Flynn, W.F.; Gao, R.; Baskin, D.S.; Pichumani, K.; Ijare, O.B.; Wood, S.L.; et al. Single-cell analysis of human glioma and immune cells identifies S100A4 as an immunotherapy target. *Nat. Commun.* **2022**, *13*, 767. [CrossRef] [PubMed]
32. Guyon, I.; Weston, J.; Barnhill, S.; Vapnik, V. Gene selection for cancer classification using support vector machines. *Mach. Learn.* **2002**, *46*, 389–422. [CrossRef]
33. Kursa, M.B. Robustness of Random Forest-based gene selection methods. *BMC Bioinform.* **2014**, *15*, 8. [CrossRef] [PubMed]
34. Zeng, D.; Ye, Z.; Shen, R.; Yu, G.; Wu, J.; Xiong, Y.; Zhou, R.; Qiu, W.; Huang, N.; Sun, L.; et al. IOBR: Multi-Omics Immuno-Oncology Biological Research to Decode Tumor Microenvironment and Signatures. *Front. Immunol.* **2021**, *12*, 687975. [CrossRef]
35. Wen, P.Y.; Kesari, S. Malignant gliomas in adults. *N. Engl. J. Med.* **2008**, *359*, 492–507. [CrossRef]
36. Nguyen, H.-M.; Guz-Montgomery, K.; Lowe, D.B.; Saha, D. Pathogenetic Features and Current Management of Glioblastoma. *Cancers* **2021**, *13*, 856. [CrossRef]
37. Jackson, C.M.; Choi, J.; Lim, M. Mechanisms of immunotherapy resistance: Lessons from glioblastoma. *Nat. Immunol.* **2019**, *20*, 1100–1109. [CrossRef]
38. Christofferson, D.E.; Yuan, J. Necroptosis as an alternative form of programmed cell death. *Curr. Opin. Cell Biol.* **2010**, *22*, 263–268. [CrossRef] [PubMed]
39. Grootjans, S.; Vanden Berghe, T.; Vandenabeele, P. Initiation and execution mechanisms of necroptosis: An overview. *Cell Death Differ.* **2017**, *24*, 1184–1195. [CrossRef]
40. Zhang, D.-W.; Shao, J.; Lin, J.; Zhang, N.; Lu, B.-J.; Lin, S.-C.; Dong, M.-Q.; Han, J. RIP3, an energy metabolism regulator that switches TNF-induced cell death from apoptosis to necrosis. *Science* **2009**, *325*, 332–336. [CrossRef]
41. Declercq, W.; Vanden Berghe, T.; Vandenabeele, P. RIP kinases at the crossroads of cell death and survival. *Cell* **2009**, *138*, 229–232. [CrossRef] [PubMed]
42. Bozec, D.; Iuga, A.C.; Roda, G.; Dahan, S.; Yeretssian, G. Critical function of the necroptosis adaptor RIPK3 in protecting from intestinal tumorigenesis. *Oncotarget* **2016**, *7*, 46384–46400. [CrossRef]

43. Feng, X.; Song, Q.; Yu, A.; Tang, H.; Peng, Z.; Wang, X. Receptor-interacting protein kinase 3 is a predictor of survival and plays a tumor suppressive role in colorectal cancer. *Neoplasma* **2015**, *62*, 592–601. [CrossRef] [PubMed]

44. Strilic, B.; Yang, L.; Albarrán-Juárez, J.; Wachsmuth, L.; Han, K.; Müller, U.C.; Pasparakis, M.; Offermanns, S. Tumour-cell-induced endothelial cell necroptosis via death receptor 6 promotes metastasis. *Nature* **2016**, *536*, 215–218. [CrossRef] [PubMed]

45. Liu, X.; Zhou, M.; Mei, L.; Ruan, J.; Hu, Q.; Peng, J.; Su, H.; Liao, H.; Liu, S.; Liu, W.; et al. Key roles of necroptotic factors in promoting tumor growth. *Oncotarget* **2016**, *7*, 22219–22233. [CrossRef] [PubMed]

46. Seifert, L.; Werba, G.; Tiwari, S.; Giao Ly, N.N.; Alothman, S.; Alqunaibit, D.; Avanzi, A.; Barilla, R.; Daley, D.; Greco, S.H.; et al. The necrosome promotes pancreatic oncogenesis via CXCL1 and Mincle-induced immune suppression. *Nature* **2016**, *532*, 245–249. [CrossRef]

47. Park, S.; Hatanpaa, K.J.; Xie, Y.; Mickey, B.E.; Madden, C.J.; Raisanen, J.M.; Ramnarain, D.B.; Xiao, G.; Saha, D.; Boothman, D.A.; et al. The receptor interacting protein 1 inhibits p53 induction through NF-kappaB activation and confers a worse prognosis in glioblastoma. *Cancer Res.* **2009**, *69*, 2809–2816. [CrossRef]

48. Zhou, Z.; Xu, J.; Huang, N.; Tang, J.; Ma, P.; Cheng, Y. Clinical and Biological Significance of a Necroptosis-Related Gene Signature in Glioma. *Front. Oncol.* **2022**, *12*, 855434. [CrossRef]

49. Hiam-Galvez, K.J.; Allen, B.M.; Spitzer, M.H. Systemic immunity in cancer. *Nat. Rev. Cancer* **2021**, *21*, 345–359. [CrossRef]

50. Wang, D.-P.; Kang, K.; Lin, Q.; Hai, J. Prognostic Significance of Preoperative Systemic Cellular Inflammatory Markers in Gliomas: A Systematic Review and Meta-Analysis. *Clin. Transl. Sci.* **2020**, *13*, 179–188. [CrossRef]

51. Han, S.; Liu, Y.; Li, Q.; Li, Z.; Hou, H.; Wu, A. Pre-treatment neutrophil-to-lymphocyte ratio is associated with neutrophil and T-cell infiltration and predicts clinical outcome in patients with glioblastoma. *BMC Cancer* **2015**, *15*, 617. [CrossRef] [PubMed]

52. Topkan, E.; Besen, A.A.; Ozdemir, Y.; Kucuk, A.; Mertsoylu, H.; Pehlivan, B.; Selek, U. Prognostic Value of Pretreatment Systemic Immune-Inflammation Index in Glioblastoma Multiforme Patients Undergoing Postneurosurgical Radiotherapy Plus Concurrent and Adjuvant Temozolomide. *Mediat. Inflamm* **2020**, *2020*, 4392189. [CrossRef]

53. Pasqualetti, F.; Giampietro, C.; Montemurro, N.; Giannini, N.; Gadducci, G.; Orlandi, P.; Natali, E.; Chiarugi, P.; Gonnelli, A.; Cantarella, M.; et al. Old and New Systemic Immune-Inflammation Indexes Are Associated with Overall Survival of Glioblastoma Patients Treated with Radio-Chemotherapy. *Genes* **2022**, *13*, 1054. [CrossRef] [PubMed]

54. Yılmaz, H.; Niğdelioğlu, B.; Oktay, E.; Meydan, N. Clinical significance of postoperatif controlling nutritional status (CONUT) score in glioblastoma multiforme. *J. Clin. Neurosci.* **2021**, *86*, 260–266. [CrossRef] [PubMed]

55. Fu, Y.; Liu, S.; Zeng, S.; Shen, H. From bench to bed: The tumor immune microenvironment and current immunotherapeutic strategies for hepatocellular carcinoma. *J. Exp. Clin. Cancer Res.* **2019**, *38*, 396. [CrossRef]

56. Schreiber, R.D.; Old, L.J.; Smyth, M.J. Cancer immunoediting: Integrating immunity's roles in cancer suppression and promotion. *Science* **2011**, *331*, 1565–1570. [CrossRef]

57. Zhang, H.; Luo, Y.-B.; Wu, W.; Zhang, L.; Wang, Z.; Dai, Z.; Feng, S.; Cao, H.; Cheng, Q.; Liu, Z. The molecular feature of macrophages in tumor immune microenvironment of glioma patients. *Comput. Struct. Biotechnol. J.* **2021**, *19*, 4603–4618. [CrossRef]

58. Zhang, C.; Zhang, M.; Song, S. Cathepsin D enhances breast cancer invasion and metastasis through promoting hepsin ubiquitin-proteasome degradation. *Cancer Lett.* **2018**, *438*, 105–115. [CrossRef]

59. Zheng, W.; Chen, Q.; Wang, C.; Yao, D.; Zhu, L.; Pan, Y.; Zhang, J.; Bai, Y.; Shao, C. Inhibition of Cathepsin D (CTSD) enhances radiosensitivity of glioblastoma cells by attenuating autophagy. *Mol. Carcinog.* **2020**, *59*, 651–660. [CrossRef]

60. Zheng, D.; Fu, W.; Jin, L.; Jiang, X.; Jiang, W.; Guan, Y.; Hao, R. The Overexpression and Clinical Significance of AP1S1 in Breast Cancer. *Cancer Manag. Res.* **2022**, *14*, 1475–1492. [CrossRef]

61. Alshabi, A.M.; Vastrad, B.; Shaikh, I.A.; Vastrad, C. Identification of Crucial Candidate Genes and Pathways in Glioblastoma Multiform by Bioinformatics Analysis. *Biomolecules* **2019**, *9*, 201. [CrossRef] [PubMed]

62. Hou, Z.; Peng, H.; White, D.E.; Wang, P.; Lieberman, P.M.; Halazonetis, T.; Rauscher, F.J. 14-3-3 binding sites in the snail protein are essential for snail-mediated transcriptional repression and epithelial-mesenchymal differentiation. *Cancer Res.* **2010**, *70*, 4385–4393. [CrossRef] [PubMed]

63. Freeman, A.K.; Morrison, D.K. 14-3-3 Proteins: Diverse functions in cell proliferation and cancer progression. *Semin. Cell Dev. Biol.* **2011**, *22*, 681–687. [CrossRef] [PubMed]

64. Aitken, A. 14-3-3 proteins: A historic overview. *Semin. Cancer Biol.* **2006**, *16*, 162–172. [CrossRef]

65. Com, E.; Clavreul, A.; Lagarrigue, M.; Michalak, S.; Menei, P.; Pineau, C. Quantitative proteomic Isotope-Coded Protein Label (ICPL) analysis reveals alteration of several functional processes in the glioblastoma. *J. Proteom.* **2012**, *75*, 3898–3913. [CrossRef] [PubMed]

66. Ni, J.; Wang, J.; Fu, Y.; Yan, C.; Zhu, M.; Jiang, Y.; Chen, J.; Ding, Y.; Fan, X.; Li, G.; et al. Functional genetic variants in centrosome-related genes CEP72 and YWHAG confer susceptibility to gastric cancer. *Arch. Toxicol.* **2020**, *94*, 2861–2872. [CrossRef] [PubMed]

67. Arlt, A.; Schäfer, H. Role of the immediate early response 3 (IER3) gene in cellular stress response, inflammation and tumorigenesis. *Eur. J. Cell Biol.* **2011**, *90*, 545–552. [CrossRef]

68. Yang, C.; Trent, S.; Ionescu-Tiba, V.; Lan, L.; Shioda, T.; Sgroi, D.; Schmidt, E.V. Identification of cyclin D1- and estrogen-regulated genes contributing to breast carcinogenesis and progression. *Cancer Res.* **2006**, *66*, 11649–11658. [CrossRef]

69. Han, L.; Geng, L.; Liu, X.; Shi, H.; He, W.; Wu, M.X. Clinical significance of IEX-1 expression in ovarian carcinoma. *Ultrastruct. Pathol.* **2011**, *35*, 260–266. [CrossRef] [PubMed]

Article

PS-NPs Induced Neurotoxic Effects in SHSY-5Y Cells via Autophagy Activation and Mitochondrial Dysfunction

Qisheng Tang [†], Tianwen Li [†], Kezhu Chen, Xiangyang Deng, Quan Zhang, Hailiang Tang, Zhifeng Shi, Tongming Zhu * and Jianhong Zhu *

Department of Neurosurgery, Huashan Hospital, Shanghai Medical College, Fudan University, National Center for Neurological Disorders, National Key Laboratory for Medical Neurobiology, Institutes of Brain Science, Shanghai Key Laboratory of Brain Function and Regeneration, MOE Frontiers Center for Brain Science, 12 Wulumuqi Zhong Rd., Shanghai 200040, China; tangqisheng@fudan.edu.cn (Q.T.); litianwen@outlook.com (T.L.); chenkzhu@163.com (K.C.); dxy623400630@163.com (X.D.); zhangquan_neuron@163.com (Q.Z.); hltang@fudan.edu.cn (H.T.); natureszf@139.com (Z.S.)
* Correspondence: ztm1911@126.com (T.Z.); jzhu@fudan.edu.cn (J.Z.)
† These authors contributed equally to this work.

Abstract: Polystyrene nanoparticles (PS-NPs) are organic pollutants that are widely detected in the environment and organisms, posing potential threats to both ecosystems and human health. PS-NPs have been proven to penetrate the blood–brain barrier and increase the incidence of neurodegenerative diseases. However, information relating to the pathogenic molecular mechanism is still unclear. This study investigated the neurotoxicity and regulatory mechanisms of PS-NPs in human neuroblastoma SHSY-5Y cells. The results show that PS-NPs caused obvious mitochondrial damages, as evidenced by inhibited cell proliferation, increased lactate dehydrogenase release, stimulated oxidative stress responses, elevated Ca^{2+} level and apoptosis, and reduced mitochondrial membrane potential and adenosine triphosphate levels. The increased release of cytochrome c and the overexpression of apoptosis-related proteins apoptotic protease activating factor-1 (Apaf-1), cysteinyl aspartate specific proteinase-3 (caspase-3), and caspase-9 indicate the activation of the mitochondrial apoptosis pathway. In addition, the upregulation of autophagy markers light chain 3-II (LC3-II), Beclin-1, and autophagy-related protein (Atg) 5/12/16L suggests that PS-NPs could promote autophagy in SHSY-5Y cells. The RNA interference of Beclin-1 confirms the regulatory role of autophagy in PS-NP-induced neurotoxicity. The administration of antioxidant N-acetylcysteine (NAC) significantly attenuated the cytotoxicity and autophagy activation induced by PS-NP exposure. Generally, PS-NPs could induce neurotoxicity in SHSY-5Y cells via autophagy activation and mitochondria dysfunction, which was modulated by mitochondrial oxidative stress. Mitochondrial damages caused by oxidative stress could potentially be involved in the pathological mechanisms for PS-NP-induced neurodegenerative diseases.

Keywords: polystyrene nanoparticles; neurotoxicity; autophagy; mitochondrial oxidative stress

Citation: Tang, Q.; Li, T.; Chen, K.; Deng, X.; Zhang, Q.; Tang, H.; Shi, Z.; Zhu, T.; Zhu, J. PS-NPs Induced Neurotoxic Effects in SHSY-5Y Cells via Autophagy Activation and Mitochondrial Dysfunction. *Brain Sci.* **2022**, *12*, 952. https://doi.org/ 10.3390/brainsci12070952

Academic Editors: James O'Callaghan and Trevor Archer

Received: 9 June 2022
Accepted: 18 July 2022
Published: 20 July 2022

1. Introduction

Polystyrene (PS) plastics are widely used as decorative materials in optical instruments and chemical departments, or they are utilized as vehicles for drugs and other types of health devices. It is estimated that several million tons of PS plastics are produced annually on a global scale. Over 80% of PS plastics are not or cannot be effectively recycled, and they are eventually released into the environment, particularly the marine environment. Discarded PS plastics can be physically broken into small pieces (nano~micro size) and then ingested by organisms; the plastics can be accumulated in the food chain due to their resistance to biodegradation [1,2]. Previous studies have found that PS nanoparticles (PS-NPs) (with a diameter < 100 nm) are globally dispersed in marine sediments, oceans, and marine organisms [3,4]. Moreover, PS-NPs have been detected in various types of human food sources, such as bottled water, tap water, fish, shrimps, beer, sugar, and sea salt, indicating a potential risk to human health [5,6]. PS-NPs may be even more hazardous than parent macroparticles due to their greater permeability across biological membranes [7]. PS-NPs are also the ideal vector for various toxic contaminants, including persistent organic pollutants, metals, pathogenic bacteria, and antibiotics [8–10]. Therefore, investigation into the potential adverse effects and molecular mechanism induced by PS-NPs is of great urgency.

The uptake of PS-NPs by organisms and the subsequent translocation to different biological tissues was first reported decades ago [11]. Studies have so far confirmed that PS-NP exposure has multiple adverse effects, such as inflammatory responses, hepatic stress, intestinal damage, reproduction disorders, nephrotoxicity, and oxidative stress stimulation [12–15]. Recent studies have found that PS-NPs can be detected in the brain tissue of *Caenorhabditis elegans*, zebrafish, and mice [16–19]. The accumulated PS-NPs in mouse brains resulted in neurobehavior alteration and cognitive impairment [20]. Penetration by PS-NPs through the blood–brain barrier (BBB) is mainly due to increased BBB permeability [21], which could potentially result from the reduced expression of tight junction proteins [22]. Mechanism research conducted by Barboza et al. indicated that the neurotoxicity induced by PS-NPs in European seabass was mediated by the induction of oxidative stress, the promotion of lipid oxidation, and the inhibition of acetylcholinesterase (AChE) activity [23]. In vitro studies have confirmed that exposure to PS-NPs (100 nm) could induce oxidative stress, increase apoptosis, and alter metabolic activity in mice primary neuronal cells [24,25]. The uptake of PS-NPs by mice hippocampal neuronal HT22 cells also induced oxidative stress and cell cycle arrest [26]. It has been documented that oxidative stress is an important mechanism for the incidence of neurodegenerative diseases following chronic long-term exposure to NPs [27,28].

The autophagy process is involved in clearing and recycling damaged proteins and cellular components [29]. It has been demonstrated that autophagy plays an essential role in the protection of neurons in the central nervous system (CNS) [30]. Considerable evidence suggests that autophagy dysregulation induced by environmental stressors is involved in the development of many neurodegenerative diseases [30,31]. A recent study proved the involvement of the autophagy pathway in embryonic toxicity, nephrotoxicity, physiological toxicity, and oxidative damage induced by NPs [32–34]. Exposure to 60 nm PS-NPs in human neuroblastoma SH-SY5Y cells also demonstrated autophagic activation, which could potentially be responsible for neural tube defects [34]. However, relatively few mechanistic studies on autophagy in potential neurotoxic effects induced by PS-NP exposure have been conducted.

A size-dependent induction by PS-NPs of autophagy initiation was observed recently in vitro in human umbilical vein endothelial cells (HUVECs) [35]. It was demonstrated that PS-NPs with a diameter of less than 100 nm could directly penetrate the cell membrane and aggregate in the cytoplasm [24,34,35]. In this study, 50 nm PS-NPs were chosen for treating human neuroblastoma SHSY-5Y cells, which is one of the most commonly used in vitro models in neurotoxicity studies [36]. Assays of cytotoxicity, autophagy activation, and mitochondrial activity were conducted. RNA interference and antioxidant N-Acetyl-L-cysteine (N-Acetylcysteine, NAC) were applied to confirm the role of oxidative stress and the autophagy process.

2. Materials and Methods

2.1. Materials

The 50 nm polystyrene nanoparticle (PS-NPs) beads (5% *w*/*v*) were purchased commercially from Janus New-Materials (Nanjing, China) (refer to the Supplementary Materials for more detailed parameters). SHSY-5Y cells, the clonal subline of neuroblastoma SK-N-SH cell line, were obtained from American Type Culture Collection (ATCC, Rockville, MD, USA). The 3-(4,5)-dimethylthiahiazo(-z-y1)-3,5-di- phenytetrazo-liumromide (MTT), Rhodamine

123 (Rh 123), 2,7-dichlorodi- hydrofluorescein diacetate (DCFH-DA), tert-Butyl hydroperoxide (tBHP), and N-Acetylcysteine (NAC) were obtained from Sigma (St. Louis, MO, USA). The annexin V-fluoresceine isothiocyanate/propidium iodide (annexin V-FITC/PI) apoptosis detection kit was obtained from Sungene (Tianjin, China). Protein extraction and assay kit was purchased from Thermo (MA, USA). Fluo 3-AM was acquired from Dojindo (Kyushu, Japan). Lactate dehydrogenase (LDH) cytotoxicity assay kit and adenosine $5'$-triphosphate (ATP) detection kit were obtained from Beyotime (Shanghai, China).

2.2. Cell Culture and PS-NP Treatment

The complete culture medium used for the SHSY-5Y cells was RPMI 1640 (Hyclone, Logan, UT, USA) containing 10% FBS (Hyclone, Logan, UT, USA), 100 U/mL penicillin, and 100 mg/L streptomycin. The SHSY-5Y cells were sub-cultured at a 10^4 cells/mL density and maintained in an incubator at 37 °C with 5% CO_2 and 100% relative humidity. Working solutions of PS-NPs were freshly diluted using the RPMI 1640 medium to 20, 50, 100, 200, and 500 mg/L. For the cytotoxicity assays, SHSY-5Y cells in the exponentially growing phase were treated with different concentrations of PS-NPs from 20 to 500 mg/L. The treated groups of 100 and 200 mg/L PS-NPs were reserved for the following mechanistic studies based on the cytotoxicity results. NAC is a sulfhydryl-containing antioxidant that can protect SHSY-5Y cells from peroxidative stress. For the NAC experiments, SHSY-5Y cells were pretreated with NAC (5 mM, 4 h) and exposed to 200 mg/L PS-NPs for 24 h. The control group cells were treated with the RPMI 1640 medium only.

2.3. Cytotoxicity Assays

SHSY-5Y cells seeded in a 96-well plate were treated with different concentrations of PS-NPs for 24 h. MTT (5 mg/mL) was then supplied, and cell viability was measured by absorbance value (570 nm) using a multifunctional microplate reader (Tecan, Männedorf, Switzerland). LDH activity was detected by the LDH assay kit with an absorbance value recorded at 490 nm. Following exposure to PS-NPs, the intracellular reactive oxygen species (ROS) level of SHSY-5Y cells seeded in a 6-well plate was measured by a fluorescent DCFH-DA assay, with excitation at 480 nm and emission at 525 nm. At the same time, the oxidative stress inducer tBHP (100 µM, 1 h) was applied as a positive control for the detection of oxidative stress [37]. The mitochondrial membrane potential (MMP, $\Delta\psi$m) was measured with a fluorescent probe Rh 123 (10 µM), with excitation at 507 nm and emission at 529 nm. The calcium ion (Ca^{2+}) content was examined by a fluorescent probe Fluo-3AM (5 µM), with excitation at 506 nm and emission at 526 nm. The fluorescence intensity was then converted to a percentage and compared to the control group [18]. The ATP levels after exposure to PS-NPs were measured with an ATP assay kit (Beyotime, Shanghai, China) according to the instructions of the manufacturer. Cell apoptosis was assayed using flow cytometry with the annexin V-FITC/PI kit according to the instructions. Detailed information related to these biological assays is provided in the Supplementary Materials.

2.4. RNA Interference (RNAi)

RNAi regulates gene expression through small interfering RNA (20–24 bp). In this study, sh-pBeclin-1 plasmid (Dharmacon, Lafayette, CO, USA) was transfected into SHSY-5Y cells using the Lipofectamine 3000 (Invitrogen, Carlsbad, CA, USA) to construct Beclin-1 RNAi (siBeclin-1) cells according to the instructions. At the same time, sh-pGIPZ plasmid (Dharmacon, Lafayette, CO, USA) was applied to construct negative control (NC) cells. NC and siBeclin-1 cells were treated with 200 mg/L of PS-NPs so that the cytotoxicity results could be verified.

2.5. Reverse Transcription Quantitative Polymerase Chain Reaction (RT-qPCR)

Following PS-NP treatment (100 and 200 mg/L, 12 h), the SHSY-5Y cells were collected and lysed using TRIZOL reagent to prepare the total RNA. The first strand of cDNA was then reverse-transcribed using a reverse transcription kit (TOYOBO, Osaka, Japan).

The cDNA was used as a template for real-time quantitative PCR amplification with the SYBR Green PCR Master Mix (Toyobo, Osaka, Japan). Glyceraldehyde-3-phosphate dehydrogenase (Gapdh) was used as an internal reference for quantifying the expression of Beclin-1. No-template controls were included in each experiment. More information, including the reaction mixture, amplification procedure, and data analysis, can be found in the Supplementary Materials.

2.6. Immunofluorescence Detection

After treatments of 100 and 200 mg/L PS-NPs for 24 h, the SHSY-5Y cells were rinsed using Hank's buffer and fixed with methanol for 10–20 min. After being washed once more using Hank's buffer, the cells were treated with Triton-X 100 (0.1%) for 10–20 min and BSA Hank's solution (1%) for 5–15 min. Then, cells were incubated with the primary antibody of cytochrome c (Cyc-c) (Abcam Company, Cambridgeshire, UK) and FITC-labeled secondary antibody (Beyotime, Shanghai, China), according to a previous study [38]. The fluorescence was excited and representative images were recorded using a fluorescent microscope (Olympus, Tokyo, Japan).

2.7. Western Blotting

After the PS-NP treatment described previously, the total protein samples of the SHSY-5Y cells were prepared using cell lysis buffer (Thermo, Waltham, MA, USA). The protein samples were subjected to sodium dodecyl sulfate-polyacrylamide gel electrophoresis and transferred to polyvinylidene fluoride membranes. The membranes were then sequentially incubated with the primary and secondary antibodies. More detailed information is provided in the Supplementary Materials. The antibodies that were used in this study included anti-light chain 3-II (LC3-II), anti-Beclin-1, anti-autophagy-related protein 12 (Atg12), anti-Atg5, anti-Atg16L, anti-Cyc-c, anti-apoptotic protease activating factor-1 (Apaf-1), anti-cysteinyl aspartate specific proteinase-3 (caspase-3), anti-caspase-9 (Cell Signaling, MA, USA), anti-GAPDH (Abcam, Cambridgeshire, UK), horseradish peroxidase (HRP)-conjugated secondary anti-mouse immunoglobulin G (IgG), and anti-rabbit IgG (Zhong Shan, China).

2.8. Statistical Analysis

Each experiment was independently repeated three times with at least triplicate replicates for each group. Data were expressed as mean $\pm$ standard deviation (SD) and analyzed by one-way analysis of variance (ANOVA). Two-way ANOVA was performed to test the interaction between the inhibitor and non-inhibitor groups. A p-value < 0.05 was considered to be statistically significant.

3. Results and Discussion

3.1. PS-NPs Induced Neurotoxicity in SHSY-5Y Cells

As an emerging organic pollutant, the potential adverse effects of plastics have attracted extensive attention, especially nanoplastics with diameters of less than 100 nm. In this study, cytotoxicity was evaluated with MTT colorimetric assay, and the viability of SHSY-5Y cells treated with PS-NPs is shown in Figure 1A. Compared to the control group, a dose-dependent inhibitory effect on cell viability was observed in cells exposed to various concentrations of PS-NPs. Low concentrations of PS-NPs (<100 mg/L) appeared to have no obvious inhibitory effects on cell viability, while exposures of 200 and 500 mg/L PS-NPs for 24 h reduced cell viability by 12.0% and 29.6%, respectively (Figure 1A). The destruction of the cell membrane results in the release of the LDH enzyme, which is widely used as a membrane integrity indicator. PS-NPs could significantly induce the release of LDH in SHSY-5Y cells. After exposure to 200 and 500 mg/L PS-NPs for 24 h, the concentrations of LDH in the culture medium increased by 39.6% and 74.3%, respectively (Figure 1B). Previous studies have also observed a significant decrease in cell viability after exposure to PS-NPs (33 nm) under comparatively high concentrations (>180 mg/L) [39], as well as obvious increases in LDH activity after PS-NP (55 nm) treatment at 250 mg/L [25].

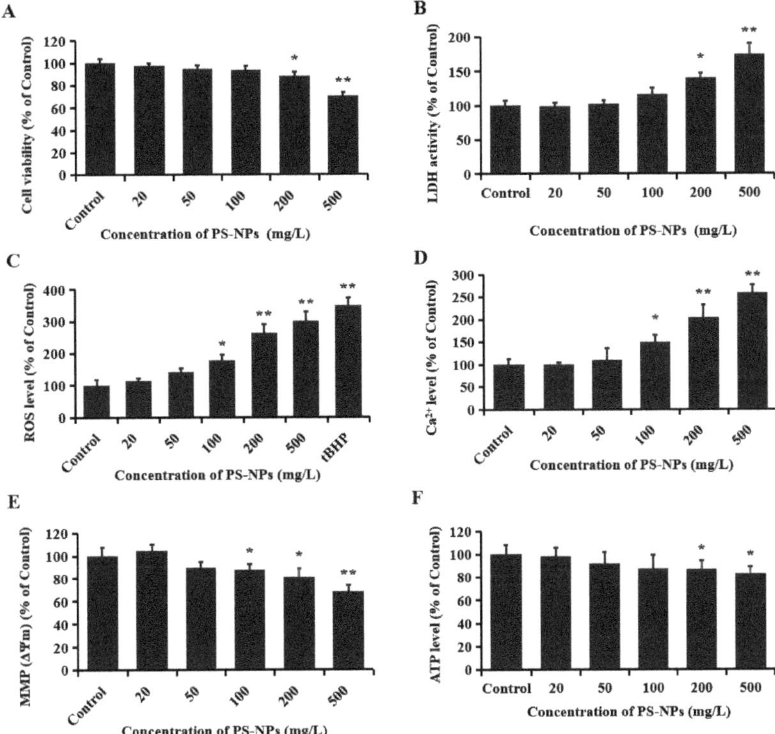

Figure 1. The cytotoxicity induced by PS-NPs in SHSY-5Y cells. SHSY-5Y cells were treated with different concentrations (20, 50, 100, 200, and 500 mg/L) of PS-NPs for 24 h. Then, cytotoxicity was measured. (**A**): Cell viability was measured with MTT. (**B**): LDH activity was detected using a LDH assay kit. (**C**): ROS production was measured using a fluorescent probe DCFH-DA, and tBHP (100 μM, 1 h) was used as a positive control for oxidative stress. (**D**): Ca^{2+} content was examined using the fluorescent probe Fluo-3AM (5 μM). (**E**): MMP ($\Delta\psi$m) was measured using the fluorescent probe Rh 123 (10 μM). (**F**): ATP level was measured using an ATP assay kit. The fluorescence intensity was converted as a percentage compared with the control. In the control group, SHSY-5Y cells were only treated with RPMI 1640 medium. * $p < 0.05$ and ** $p < 0.01$, compared to the control.

3.2. PS-NPs Promoted Oxidative Stress in SHSY-5Y Cells

The byproducts of mitochondrial metabolism include the buildup of reactive oxygen species (ROS) and the concentration of calcium ion (Ca^{2+}) [40]. Excessive production of ROS can cause oxidative damage, disrupt homeostasis, and significantly suppress cell viability. Oxidative stress has been confirmed by previous studies to be a prominent neurotoxic mechanism after exposure to exogenous toxicants [41,42]. The activation of oxidative stress is closely associated with the onset of neurodegenerative diseases [43]. As shown in Figure 1C, PS-NP treatment at concentrations of >100 mg/L could obviously increase intracellular ROS production. There were 1.8-, 2.6-, and 3.0-fold increases in the ROS levels in the 100, 200, or 500 mg/L PS-NP treatment groups, respectively. An earlier study discovered that PS-NPs could significantly enter human cerebral microvascular endothelial hCMEC/D3 cells and mouse hippocampal neuronal HT22 cells, which resulted in a significant increase in ROS production [21,26].

3.3. PS-NPs Induced Mitochondrial Damage in SHSY-5Y Cells

As a secondary messenger, calcium ions (Ca^{2+}) are essential for maintaining physiological nerve conduction function. Mitochondria take part in the regulation of the intracellular Ca^{2+} levels, functioning as a local Ca^{2+} buffer pool. The entry of Ca^{2+} into mitochondria is driven by the mitochondrial inner membrane ($\Delta\Psi m$) electrochemical gradient. Ca^{2+} accumulation and the consequent mitochondrial permeability transition pore (mPTP) opening can lead to the dissipation of $\Delta\Psi m$ and ultimately induce the cell death signaling pathways [44]. In this study, the fluorescent probes Fluo-3AM and Rh123 were used to detect the Ca^{2+} concentrations and MMP in SHSY-5Y cells. Our results show that the Ca^{2+} level was markedly increased and the mitochondrial membrane potential (MMP, $\Delta\Psi m$) was significantly reduced after treatment with 100, 200, and 500 mg/L PS-NP (Figure 1D,E), which indicates the induction of mitochondrial damage. These results are in agreement with previous findings that 55 nm PS-NPs influenced the mitochondrial activity of neural cells at high concentrations (250 mg/L) [25]. Other particles, such as combustion-derived particles, could also affect intracellular calcium homeostasis, contributing to the development or aggravation of cardiovascular disease [45].

Adenosine triphosphate (ATP) participates in various physiological processes in organisms, functioning as an active energy molecule, which is produced in the folds of the inner membrane of mitochondria. When the mitochondria are damaged, the switching state of the mitochondrial membrane voltage-dependent anion channel (VDAC) is affected, thereby inhibiting the release of macromolecule anionic ATP [46]. Especially in the state of apoptosis or necrosis, a decrease in ATP levels generally occurs simultaneously with the loss of MMP [47]. In comparison to the control group, the ATP levels in the SHSY-5Y cells were reduced by 13.2%, and 17.3%, respectively, after treatment with 200 and 500 mg/L PS-NP (Figure 1F). This finding further demonstrates the mitochondrial damage in SHSY-5Y cells induced by PS-NPs. All these data are consistent with the results of cell viability, suggesting that oxidative stress and mitochondrial respiration defect might be involved in the cytotoxicity induced by PS-NPs.

3.4. PS-NPs Induced Mitochondrial Apoptosis in SHSY-5Y Cells

Apoptosis is a complicated process in multicellular organisms to remove abnormal cells and maintain cellular homeostasis, which is regulated by complex signaling pathways [48]. The imbalance of apoptotic homeostasis is closely associated with numerous diseases, such as cancer, autoimmune diseases, neurological disorders, and cardiovascular disorders [49]. Oxidative stress has been proven to be one of the primary reasons for the induction of apoptosis [50]. To investigate whether the mitochondrial apoptotic pathway is involved in the cytotoxicity of PS-NPs in SHSY-5Y cells, cell apoptosis was detected using the annexin V-FITC/PI kit. Our results show a positive correlation between the apoptosis ratio and PS-NP concentration, with significant apoptosis induction at doses of 100, 200, and 500 mg/L (Figure 2A). In mammal cells, mitochondrial damage can cause the release of cytochrome c (Cyc-c) into the cytoplasm, forming a complex with cysteinyl aspartate specific proteinase-9 (caspase-9) and apoptotic protease activating factor-1 (Apaf-1), which can activate caspase-3, and ultimately result in apoptosis [51]. The localization transition of Cyc-c in cells was monitored in this study using an immunofluorescence assay. Based on the results of cytotoxicity, moderate doses of 100 and 200 mg/L were used for subsequent mechanistic studies. As shown in Figure 2B, the aggregation of Cyc-c in the cytoplasm was observed, indicating a release of Cyc-c from the mitochondria into the cytoplasm. The results of Western blotting show that the protein expressions of caspase-3 increased to 123.6% and 178.5% in the control group, after treatment with 100 and 200 mg/L PS-NPs. Synchronously, the expressions of caspase-9 and Aparf-1 show similar increasing trends to 136.2% and 185.3%, and 121.3% and 159.4% (Figure 2C,D), respectively. These results provide evidence that PS-NPs could trigger oxidative stress as an early response, eventually resulting in apoptosis via the mitochondrial apoptotic pathway.

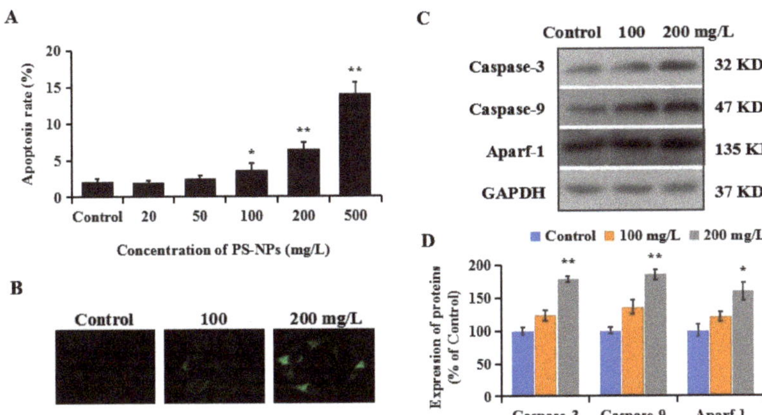

Figure 2. The apoptosis induced by PS-NPs in SHSY-5Y cells. (**A**): SHSY-5Y cells were treated with different concentrations (20, 50, 100, 200, and 500 mg/L) of PS-NPs for 24 h. Then, apoptosis was detected with the annexin V-FITC/PI kit. (**B**): After the treatment of 100 and 200 mg/L of PS-NPs for 24 h, the localization of Cyc-c was observed by immunofluorescence. (**C**): The expressions of caspase-3, caspase-9, and Aparf-1 were measured with Western blotting. (**D**): The expressions of proteins were quantified. In the control group, SHSY-5Y cells were only treated with medium. * $p < 0.05$ and ** $p < 0.01$, compared to the control.

3.5. PS-NPs Activated Autophagy in SHSY-5Y Cells

Autophagy is a conserved phylogenetic process, playing a vital role in the elimination of aggregated proteins and damaged organelles [29]. A dysregulated autophagy process is related to the development of neurodegenerative diseases [52]. Autophagic initiation is a complex multi-step and crosstalk process that could be induced by an alteration in redox signaling [53]. Excessive ROS is a well-known inducer of autophagy initiation [54]. Our results show that PS-NPs have the ability to induce oxidative stress and mitochondria dysfunction. To investigate the autophagy status induced by PS-NPs in SHSY-5Y cells, the expression levels of LC3-II were measured using Western blotting. LC3 is a homolog of autophagy-related protein-8 (Atg8) in mammalian cells. LC3-I (located in the cytoplasm) is covalently connected with phosphatidylethanolamine and inserted into the bilayer membrane of autophagy to form LC3-II. The conversion of LC3-I to LC3-II has been used as a recognized indicator for the evaluation and determination of autophagy [29]. As depicted in Figure 3A,B, a significant increase in LC3-II expression was observed after PS-NP exposure (Figure 3A), suggesting autophagy initiation in SHSY-5Y cells treated with PS-NPs.

To confirm the formation of autophagy in the SHSY-5Y cells, the expressions of Beclin-1 and Atg5/12/16L1 were measured using Western blotting. Beclin-1 is a subunit of the phosphatidylinositol 3 kinase complex. The binding of Beclin-1 to autophagy precursors results in the initiation of autophagosome formation [55]. The triplet complex that is formed by the sequential conjugation of Atg12, Atg5, and Atg16L is indispensable for autophagosome formation [56]. Our results show that the levels of Beclin-1 were elevated in SHSY-5Y cells exposed to PS-NPs compared to the control (Figure 3A,B). The expressions of Beclin-1 increased to 132.6% and 178.3%, respectively, after exposure to 100 and 200 mg/L PS-NPs. The expressions of Atg12, Atg5, and Atg16L show similar increasing changes, with inductions of 32.1–113.0% (Figure 3A,B). These results further confirm the induction of autophagy by PS-NPs. Wang et al. reported that human kidney proximal tubular epithelial HK-2 cells had higher protein levels of LC3-II and Beclin-1 after exposure to PS-MPs (2 μm) [57]. The data of Nie et al. indicated the upregulation of Atg7, Atg5, and LC3-II protein expression and the fluorescence intensity of GFP-LC3 in SH-SY5Y cells after exposure to 60 nm PS-NPs [34]. Exposure to 100 nm PS-NPs could increase LC3 autophagic pathway activation in mouse embryonic fibroblasts (MEFs) [58].

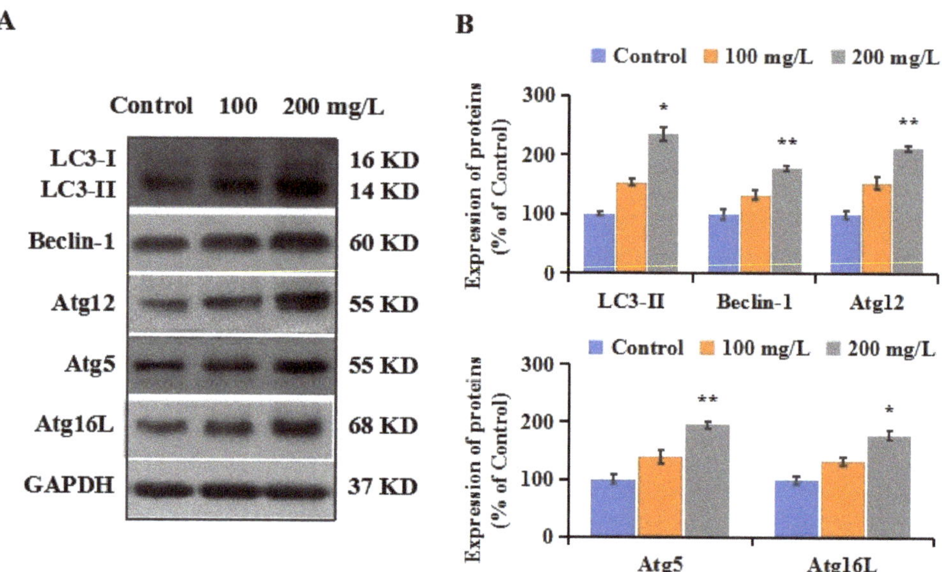

Figure 3. The autophagy induced by PS-NPs in SHSY-5Y cells. After the treatment of 100 and 200 mg/L of PS-NPs for 24 h. (**A**): The expressions of autophagy-related proteins including LC3-II, Beclin-1, Atg12, Atg5, and Atg16L were measured with Western blotting. (**B**): The expressions of proteins were quantified. In the control group, SHSY-5Y cells were only treated with medium. * $p < 0.05$ and ** $p < 0.01$, compared to the control.

3.6. Regulatory Role of Autophagy in PS-NP-Induced Neurotoxicity

To verify the involvement of autophagy in the neurotoxicity induced by PS-NPs in SHSY-5Y cells, autophagy-blocked RNAi was conducted. Sh-pBeclin-1 and sh-pGIPZ plasmids were transfected into SHSY-5Y cells as described in the method section, and then the gene levels and protein expressions of Beclin-1 were analyzed using RT-qPCR and Western blotting, respectively. The results show that gene expression decreased by 76.5%, and protein expression decreased by 58.4% in siBeclin-1 SHSY-5Y cells compared to the control, indicating that autophagy was effectively knocked down in the initial step (Figure 4A). As displayed in Figure 4B, the upregulation of LC3-II induced by PS-NPs was obviously suppressed compared to NC cells. Similarly, in a study on sevoflurane-induced neurotoxicity, a significant decrease in LC3-II expression was observed after the transfection of siRNA Beclin-1 in neonatal rat hippocampal cells [59]. At the same time, biological influences on SHSY-5Y cells (including LDH release, ROS level, MMP, and apoptosis rate) were significantly relieved in siBeclin-1 SHSY-5Y cells. As shown in Figure 4C, the activity of LDH after exposure to 200 mg/L PS-NPs decreased by 14.5% in the siBeclin-1 cell group, and the production of ROS and the apoptosis rate were reduced by over 30%. In addition, the depolarization of MMP was also alleviated in the siBeclin-1 group. Previous studies have also reported that the inhibition of autophagy by Beclin-1 siRNA in neonatal rat hippocampal cells reduced sevoflurane-induced apoptosis [59]. The downregulation of Beclin-1 attenuated viral-induced inflammation in HIV-1-infected microglial cells [60]. These results indicate that autophagy is involved in the regulation of neurotoxicity induced by PS-NPs, which is also in accordance with the study conducted by Yin et al., showing that microplastics triggered autophagy-dependent ferroptosis and apoptosis in chicken brains [22].

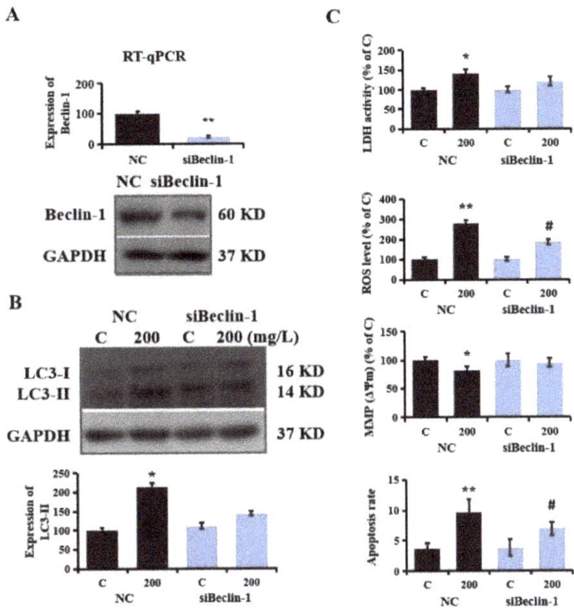

Figure 4. Autophagy-regulated neurotoxicity induced by PS-NPs. (**A**): SHSY-5Y cells were transfected with sh-pBeclin-1 and sh-pGIPZ plasmids to construct Beclin-1 RNAi cells (siBeclin-1) and negative cells (NC). The expression of Beclin-1 was analyzed with RT-qPCR and Western blotting. NC and siBeclin-1 cells were treated with 200 mg/L of PS-NPs, and then the expressions of LC3-II were measured with Western blotting, and the expressions of proteins were quantified (**B**). The LDH activity, ROS level, MMP, and apoptosis rate were measured (**C**). In the C (control) group, SHSY-5Y cells were only treated with medium. * $p < 0.05$ and ** $p < 0.01$, compared to the control. # $p < 0.05$, siBeclin-1 group compared to the NC group.

3.7. PS-NP-Induced Neurotoxicity Attenuated by Oxidative Antioxidant

Accumulating evidence indicates the central role of autophagy in the mammalian oxidative stress response. The dysregulation of redox homeostasis can activate autophagy to degrade or recycle intracellular damaged macromolecules and facilitate cellular adaptation [61]. The antioxidant NAC can antagonize arsenic-induced neurotoxicity through the suppression of oxidative stress in mouse oligodendrocyte precursor cells [62]. The addition of NAC can abolish H_2O_2-induced autophagy and inhibit mitochondrial ROS production [63]. To confirm the role of oxidative stress in PS-NP-induced autophagy initiation and neurotoxicity, NAC was applied in the present study. NAC pretreatment obviously reduced ROS generation induced by PS-NPs in SHSY-5Y cells (Figure 5A). Similarly, LDH release, MMP reduction, and apoptosis induction in PS-NP-treated SHSY-5Y cells were significantly attenuated by the administration of NAC (Figure 5A). Additionally, NAC alleviated mitochondrial apoptosis in SHSY-5Y cells, demonstrated by reduced caspase-3, caspase-9, and Aparf-1 protein levels (Figure 5B). Moreover, the expressions of LC3-II and Beclin-1 expression in SHSY-5Y cells exposed to PS-NPs were significantly reduced by NAC pretreatment (Figure 5B). Liu et al. reported that the suppression of oxidative stress by NAC rescued the inhibition of cell growth and induction of autophagy caused by plasticizer tri-ortho-cresyl phosphate (TOCP) treatment in rat spermatogonia stem cells [64]. The pretreatment of antioxidant nanomaterial poly-amidoamine effectively impaired the autophagic effects and reduced neuronal cell death [65]. Our data confirm that mitochondrial oxidative stress plays a vital role in neurotoxicity and autophagy initiation induced by PS-NPs in SHSY-5Y cells, indicating that mitochondrial oxidative stress is a regulatory mechanism of neurotoxicity.

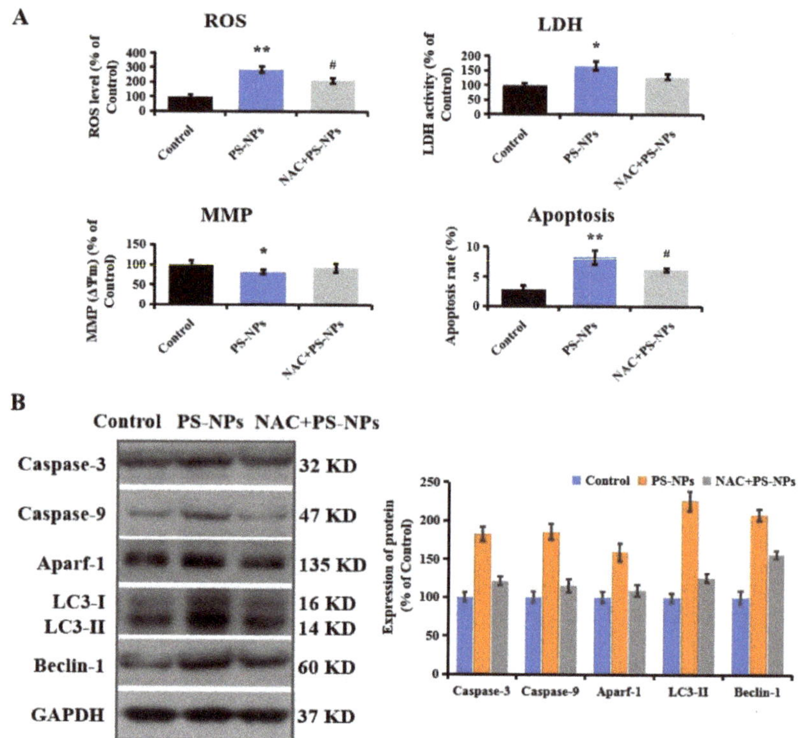

Figure 5. Oxidative stress regulated autophagy and neurotoxicity induced by PS-NPs. SHSY-5Y cells were pretreated with NAC (5 mM) for 4 h and then treated with 200 mg/L of PS-NPs for 24 h. (**A**): The ROS level, LDH activity, MMP, and apoptosis rate were measured. (**B**): The expression of apoptosis and autophagy-related proteins were measured with Western blotting, and the expressions of proteins were quantified. In the control group, SHSY-5Y cells were only treated with medium. * $p < 0.05$ and ** $p < 0.01$, compared to the control. # $p < 0.05$, NAC+PS-NPs group compared to the PS-NPs group.

4. Conclusions

In conclusion, exposure to PS-NPs reduced cell viability and induced the release of LDH in SHSY-5Y cells in a concentration-dependent manner. PS-NPs could significantly induce oxidative stress and mitochondrial dysfunction as indicated by increased intracellular ROS production and Ca^{2+} concentrations and decreased MMP and ATP levels. Moreover, PS-NPs activated the mitochondrial apoptotic pathway and promoted cell apoptosis. The upregulation of autophagy marker proteins indicates the stimulation of the autophagy process after PS-NP exposure. The suppression of Beclin-1 with RNAi revealed the regulatory role of autophagy in neurotoxicity induced by PS-NPs. The results of antioxidant NAC suggest that mitochondrial oxidative stress could regulate PS-NP-induced neurotoxicity by the modulation of autophagy activation. In general, high concentrations (>100 mg/L) of PS-NPs resulted in significant neurotoxic effects by the activation of autophagy and mitochondrial dysfunction, which was modulated by oxidative stress. The stimulation of mitochondrial oxidative stress and autophagy suggests the potential pathological mechanisms of neurodegenerative diseases induced by PS-NPs.

Supplementary Materials: The supplementary material associated with this article can be found online: https://www.mdpi.com/article/10.3390/brainsci12070952/s1. The detailed parameters of PS-NPs. Detailed information about MTT assay, LDH release assay, ROS measure, MMP ($\Delta\psi$m) measure, intracellular calcium ion (Ca^{2+}) content measure, ATP level measure, apoptosis assay, reverse transcription quantitative polymerase chain reaction (RT-qPCR), and Western blotting. The flow cytometry dot plot of apoptosis. The large-sized images of immunofluorescence. The effect of NAC and tBHP on cell viability. Full image of Western blotting.

Author Contributions: Q.T. and T.L. performed the majority of the investigation, and it is considered that their contribution was similar. K.C. participated in data formal analysis. X.D. participated in data curation. Q.Z. took part in methodology. H.T. and Z.S. wrote the first draft of the manuscript. J.Z. and T.Z. participated in the project administration, supervision, and writing—review and editing. All authors have read and agreed to the published version of the manuscript.

Funding: This work was supported by grants from the Ministry of Science and Technology of China (2018YFA0107900), National Nature Science Foundation (92168103, 32171417), and Shanghai Municipal Government (2019CXJQ01), Peak Disciplines (Type IV) of Institutions of Higher Leaning in Shanghai.

Institutional Review Board Statement: Not applicable.

Informed Consent Statement: Not applicable.

Data Availability Statement: Data are contained within the article or supplementary material.

References

1. Barnes, D.K.; Galgani, F.; Thompson, R.C.; Barlaz, M. Accumulation and fragmentation of plastic debris in global environments. *Philos. Trans. R Soc. Lond. B Biol. Sci.* **2009**, *364*, 1985–1998. [CrossRef] [PubMed]
2. Wright, S.L.; Thompson, R.C.; Galloway, T.S. The physical impacts of microplastics on marine organisms: A review. *Environ. Pollut.* **2013**, *178*, 483–492. [CrossRef]
3. Lehner, R.; Weder, C.; Petri-Fink, A.; Rothen-Rutishauser, B. Emergence of nanoplastic in the environment and possible impact on human health. *Environ. Sci. Technol.* **2019**, *53*, 1748–1765. [CrossRef]
4. Martin, J.; Lusher, A.; Thompson, R.C.; Morley, A. The deposition and accumulation of microplastics in marine sediments and bottom water from the irish continental shelf. *Sci. Rep.* **2017**, *7*, 10772. [CrossRef]
5. Bouwmeester, H.; Hollman, P.C.; Peters, R.J. Potential health impact of environmentally released micro- and nanoplastics in the human food production chain: Experiences from nanotoxicology. *Environ. Sci. Technol.* **2015**, *49*, 8932–8947. [CrossRef] [PubMed]
6. Barboza, L.G.A.; Dick Vethaak, A.; Lavorante, B.R.B.O.; Lundebye, A.K.; Guilhermino, L. Marine microplastic debris: An emerging issue for food security, food safety and human health. *Mar. Pollut. Bull.* **2018**, *133*, 336–348. [CrossRef] [PubMed]
7. Chen, Q.; Gundlach, M.; Yang, S.; Jiang, J.; Velki, M.; Yin, D.; Hollert, H. Quantitative investigation of the mechanisms of microplastics and nanoplastics toward zebrafish larvae locomotor activity. *Sci. Total Environ.* **2017**, *584–585*, 1022–1031. [CrossRef]
8. Koelmans, A.A.; Bakir, A.; Burton, G.A.; Janssen, C.R. Microplastic as a vector for chemicals in the aquatic environment: Critical review and model-supported reinterpretation of empirical studies. *Environ. Sci. Technol.* **2016**, *50*, 3315–3326. [CrossRef]
9. Koelmans, A.A.; Besseling, E.; Wegner, A.; Foekema, E.M. Plastic as a carrier of POPs to aquatic organisms: A model analysis. *Environ. Sci. Technol.* **2013**, *47*, 7812–7820. [CrossRef]
10. Li, J.; Zhang, K.; Zhang, H. Adsorption of antibiotics on microplastics. *Environ. Pollut.* **2018**, *237*, 460–467. [CrossRef]
11. Jani, P.; Halbert, G.W.; Langridge, J.; Florence, A.T. Nanoparticle uptake by the rat gastrointestinal mucosa: Quantitation and particle size dependency. *J. Pharm. Pharmacol.* **1990**, *42*, 821–826. [CrossRef] [PubMed]
12. Ferreira, P.; Fonte, E.; Soares, M.E.; Carvalho, F.; Guilhermino, L. Effects of multi-stressors on juveniles of the marine fish Pomatoschistus microps: Gold nanoparticles, microplastics and temperature. *Aquat. Toxicol.* **2016**, *170*, 89–103. [CrossRef] [PubMed]
13. Pedà, C.; Caccamo, L.; Fossi, M.C.; Gai, F.; Andaloro, F.; Genovese, L.; Perdichizzi, A.; Romeo, T.; Maricchiolo, G. Intestinal alterations in European sea bass *Dicentrarchus labrax* (Linnaeus, 1758) exposed to microplastics: Preliminary results. *Environ. Pollut.* **2016**, *212*, 251–256. [CrossRef]
14. Rochman, C.M.; Hoh, E.; Kurobe, T.; The, S.J. Ingested plastic transfers hazardous chemicals to fish and induces hepatic stress. *Sci. Rep.* **2013**, *3*, 3263. [CrossRef] [PubMed]
15. von Moos, N.; Burkhardt-Holm, P.; Köhler, A. Uptake and effects of microplastics on cells and tissue of the blue mussel *Mytilus edulis* L. after an experimental exposure. *Environ. Sci. Technol.* **2012**, *46*, 11327–11335. [CrossRef]
16. Prüst, M.; Meijer, J.; Westerink, R.H.S. The plastic brain: Neurotoxicity of micro- and nanoplastics. *Part. Fibre Toxicol.* **2020**, *17*, 24. [CrossRef]

17. Sarasamma, S.; Audira, G.; Siregar, P.; Malhotra, N.; Lai, Y.H.; Liang, S.T.; Chen, J.R.; Chen, K.H.; Hsiao, C.D. Nanoplastics cause neurobehavioral impairments, reproductive and oxidative damages, and biomarker responses in Zebrafish: Throwing up alarms of wide spread health risk of exposure. *Int. J. Mol. Sci.* **2020**, *21*, 1410. [CrossRef]

18. Shang, Y.; Wang, S.; Jin, Y.; Xue, W.; Zhong, Y.; Wang, H.; An, J.; Li, H. Polystyrene nanoparticles induced neurodevelopmental toxicity in *Caenorhabditis elegans* through regulation of dpy-5 and rol-6. *Ecotoxicol. Environ. Saf.* **2021**, *222*, 112523. [CrossRef]

19. Liu, X.; Zhao, Y.; Dou, J.; Hou, Q.; Cheng, J.; Jiang, X. Bioeffects of inhaled nanoplastics on neurons and alteration of animal behaviors through beposition in the brain. *Nano Lett.* **2022**, *22*, 1091–1099. [CrossRef]

20. Estrela, F.N.; Guimarães, A.; Araújo, A.; Silva, F.G.; Luz, T.; Silva, A.M.; Pereira, P.S.; Malafaia, G. Toxicity of polystyrene nanoplastics and zinc oxide to mice. *Chemosphere* **2021**, *271*, 129476. [CrossRef]

21. Shan, S.; Zhang, Y.; Zhao, H.; Zeng, T.; Zhao, X. Polystyrene nanoplastics penetrate across the blood-brain barrier and induce activation of microglia in the brain of mice. *Chemosphere* **2022**, *298*, 134261. [CrossRef] [PubMed]

22. Yin, K.; Wang, D.; Zhao, H.; Wang, Y.; Zhang, Y.; Liu, Y.; Li, B.; Xing, M. Polystyrene microplastics up-regulates liver glutamine and glutamate synthesis and promotes autophagy-dependent ferroptosis and apoptosis in the cerebellum through the liver-brain axis. *Environ. Pollut.* **2022**, *307*, 119449. [CrossRef] [PubMed]

23. Barboza, L.G.A.; Vieira, L.R.; Branco, V.; Figueiredo, N.; Carvalho, F.; Carvalho, C.; Guilhermino, L. Microplastics cause neurotoxicity, oxidative damage and energy-related changes and interact with the bioaccumulation of mercury in the European seabass, *Dicentrarchus labrax* (Linnaeus, 1758). *Aquat. Toxicol.* **2018**, *195*, 49–57. [CrossRef] [PubMed]

24. Jung, B.K.; Han, S.W.; Park, S.H.; Bae, J.S.; Choi, J.; Ryu, K.Y. Neurotoxic potential of polystyrene nanoplastics in primary cells originating from mouse brain. *Neurotoxicology* **2020**, *81*, 189–196. [CrossRef]

25. Murali, K.; Kenesei, K.; Li, Y.; Demeter, K.; Környei, Z.; Madarász, E. Uptake and bio-reactivity of polystyrene nanoparticles is affected by surface modifications, ageing and LPS adsorption: In vitro studies on neural tissue cells. *Nanoscale* **2015**, *7*, 4199–4210. [CrossRef]

26. Liu, S.; Li, Y.; Shang, L.; Yin, J.; Qian, Z.; Chen, C.; Yang, Y. Size-dependent neurotoxicity of micro- and nanoplastics in flowing condition based on an in vitro microfluidic study. *Chemosphere* **2022**, *303 Pt 3*, 135280. [CrossRef]

27. Migliore, L.; Coppedè, F. Environmental-induced oxidative stress in neurodegenerative disorders and aging. *Mutat. Res.* **2009**, *674*, 73–84. [CrossRef]

28. Fröhlich, E.; Meindl, C.; Roblegg, E.; Ebner, B.; Absenger, M.; Pieber, T.R. Action of polystyrene nanoparticles of different sizes on lysosomal function and integrity. *Part. Fibre Toxicol.* **2012**, *9*, 26. [CrossRef]

29. Galluzzi, L.; Bravo-San Pedro, J.M.; Blomgren, K.; Kroemer, G. Autophagy in acute brain injury. *Nat. Rev. Neurosci.* **2016**, *17*, 467–484. [CrossRef]

30. Luo, J. Autophagy and ethanol neurotoxicity. *Autophagy* **2014**, *10*, 2099–2108. [CrossRef]

31. Giordano, S.; Darley-Usmar, V.; Zhang, J. Autophagy as an essential cellular antioxidant pathway in neurodegenerative disease. *Redox Biol.* **2013**, *2*, 82–90. [CrossRef] [PubMed]

32. Chen, W.; Chu, Q.; Ye, X.; Sun, Y.; Liu, Y.; Jia, R.; Li, Y.; Tu, P.; Tang, Q.; Yu, T.; et al. Canidin-3-glucoside prevents nano-plastics induced toxicity via activating autophagy and promoting discharge. *Environ. Pollut.* **2021**, *274*, 116524. [CrossRef] [PubMed]

33. Chen, Y.C.; Chen, K.F.; Lin, K.A.; Chen, J.K.; Jiang, X.Y.; Lin, C.H. The nephrotoxic potential of polystyrene microplastics at realistic environmental concentrations. *J. Hazard Mater.* **2022**, *427*, 127871. [CrossRef]

34. Nie, J.H.; Shen, Y.; Roshdy, M.; Cheng, X.; Wang, G.; Yang, X. Polystyrene nanoplastics exposure caused defective neural tube morphogenesis through caveolae-mediated endocytosis and faulty apoptosis. *Nanotoxicology* **2021**, *15*, 885–904. [CrossRef] [PubMed]

35. Lu, Y.Y.; Li, H.; Ren, H.; Zhang, X.; Huang, F.; Zhang, D.; Huang, Q.; Zhang, X. Size-dependent effects of polystyrene nanoplastics on autophagy response in human umbilical vein endothelial cells. *J. Hazard Mater.* **2022**, *421*, 126770. [CrossRef] [PubMed]

36. Kovalevich, J.; Langford, D. Considerations for the use of SH-SY5Y neuroblastoma cells in neurobiology. *Methods Mol. Biol.* **2013**, *1078*, 9–21.

37. Zhou, J.; Liu, Q.; Yang, Z.; Xie, C.; Ling, L.; Hu, H.; Cao, Y.; Huang, Y.; Hua, Y. Rutin maintains redox balance to relieve oxidative stress induced by TBHP in nucleus pulposus cells. *In Vitro Cell. Dev. Biol. Anim.* **2021**, *57*, 448–456. [CrossRef] [PubMed]

38. Nur-E-Kamal, A.; Gross, S.R.; Pan, Z.; Balklava, Z.; Ma, J.; Liu, L.F. Nuclear translocation of cytochrome c during apoptosis. *J. Biol. Chem.* **2004**, *279*, 24911–24914. [CrossRef]

39. Hoelting, L.; Scheinhardt, B.; Bondarenko, O.; Schildknecht, S.; Kapitza, M.; Tanavde, V.; Tan, B.; Lee, Q.Y.; Mecking, S.; Leist, M.; et al. A 3-dimensional human embryonic stem cell (hESC)-derived model to detect developmental neurotoxicity of nanoparticles. *Arch. Toxicol.* **2013**, *87*, 721–733. [CrossRef]

40. Zorov, D.B.; Juhaszova, M.; Sollott, S.J. Mitochondrial reactive oxygen species (ROS) and ROS-induced ROS release. *Physiol. Rev.* **2014**, *94*, 909–950. [CrossRef]

41. Park, J.C.; Han, J.; Lee, M.C.; Seo, J.S.; Lee, J.S. Effects of triclosan (TCS) on fecundity, the antioxidant system, and oxidative stress-mediated gene expression in the copepod *Tigriopus japonicus*. *Aquat. Toxicol.* **2017**, *189*, 16–24. [CrossRef] [PubMed]

42. Wang, S.C.; Gao, Z.Y.; Liu, F.F.; Chen, S.Q.; Liu, G.Z. Effects of polystyrene and triphenyl phosphate on growth, photosynthesis and oxidative stress of Chaetoceros meülleri. *Sci. Total Environ.* **2021**, *797*, 149180. [CrossRef] [PubMed]

43. Angelova, P.R.; Abramov, A.Y. Role of mitochondrial ROS in the brain: From physiology to neurodegeneration. *FEBS Lett.* **2018**, *592*, 692–702. [CrossRef]

44. McKenzie, M.; Lim, S.C.; Duchen, M.R. Simultaneous measurement of mitochondrial calcium and mitochondrial membrane potential in live cells by fluorescent microscopy. *J. Vis. Exp.* **2017**, *119*, 55166. [CrossRef]

45. Holme, J.A.; Brinchmann, B.C.; Le Ferrec, E.; Lagadic-Gossmann, D.; Øvrevik, J. Combustion Particle-Induced Changes in Calcium Homeostasis: A Contributing Factor to Vascular Disease? *Cardiovasc. Toxicol.* **2019**, *19*, 198–209. [CrossRef] [PubMed]

46. Mazure, N.M. VDAC in cancer. *Biochim. Biophys. Acta Bioenerg.* **2017**, *1858*, 665–673. [CrossRef] [PubMed]

47. Su, X.; Lv, L.; Li, Y.; Fang, R.; Yang, R.; Li, C.; Li, T.; Zhu, D.; Li, X.; Zhou, Y.; et al. lncRNA MIRF promotes cardiac apoptosis through the miR-26a-Bak1 axis. *Mol. Ther. Nucleic Acids* **2020**, *20*, 841–850. [CrossRef]

48. McComb, S.; Chan, P.K.; Guinot, A.; Hartmannsdottir, H.; Jenni, S.; Dobay, M.P.; Bourquin, J.P.; Bornhauser, B.C. Efficient apoptosis requires feedback amplification of upstream apoptotic signals by effector caspase-3 or -7. *Sci. Adv.* **2019**, *5*, eaau9433. [CrossRef]

49. Favaloro, B.; Allocati, N.; Graziano, V.; Di Ilio, C.; De Laurenzi, V. Role of apoptosis in disease. *Aging* **2012**, *4*, 330–349. [CrossRef]

50. Redza-Dutordoir, M.; Averill-Bates, D.A. Activation of apoptosis signalling pathways by reactive oxygen species. *Biochim. Biophys. Acta* **2016**, *1863*, 2977–2992. [CrossRef]

51. Van Opdenbosch, N.; Lamkanfi, M. Caspases in cell death, inflammation, and disease. *Immunity* **2019**, *50*, 1352–1364. [CrossRef] [PubMed]

52. Ariosa, A.R.; Klionsky, D.J. Autophagy core machinery: Overcoming spatial barriers in neurons. *J. Mol. Med.* **2016**, *94*, 1217–1227. [CrossRef] [PubMed]

53. Yun, H.R.; Jo, Y.H.; Kim, J.; Shin, Y.; Kim, S.S.; Choi, T.G. Roles of autophagy in oxidative stress. *Int. J. Mol. Sci.* **2020**, *21*, 3289. [CrossRef] [PubMed]

54. Scherz-Shouval, R.; Elazar, Z. Regulation of autophagy by ROS: Physiology and pathology. *Trends Biochem. Sci.* **2011**, *36*, 30–38. [CrossRef]

55. Maejima, Y.; Isobe, M.; Sadoshima, J. Regulation of autophagy by Beclin 1 in the heart. *J. Mol. Cell. Cardiol.* **2016**, *95*, 19–25. [CrossRef]

56. Hamaoui, D.; Subtil, A. ATG16L1 functions in cell homeostasis beyond autophagy. *FEBS J.* **2022**, *289*, 1779–1800. [CrossRef]

57. Wang, Y.L.; Lee, Y.H.; Hsu, Y.H.; Chiu, I.J.; Huang, C.C.; Huang, C.C.; Chia, Z.C.; Lee, C.P.; Lin, Y.F.; Chiu, H.W. The kidney-related effects of polystyrene microplastics on human kidney proximal tubular epithelial cells HK-2 and male C57BL/6 mice. *Environ. Health Perspect.* **2021**, *129*, 57003. [CrossRef]

58. Han, S.W.; Choi, J.; Ryu, K.Y. Stress response of mouse embryonic fibroblasts exposed to polystyrene nanoplastics. *Int. J. Mol. Sci.* **2021**, *22*, 2094. [CrossRef]

59. Xu, L.; Shen, J.; Yu, L.; Sun, J.; McQuillan, P.M.; Hu, Z.; Yan, M. Role of autophagy in sevoflurane-induced neurotoxicity in neonatal rat hippocampal cells. *Brain Res. Bull.* **2018**, *140*, 291–298. [CrossRef]

60. Rodriguez, M.; Kaushik, A.; Lapierre, J.; Dever, S.M.; El-Hage, N.; Nair, M. Electro-Magnetic Nano-particle bound Beclin1 siRNA crosses the blood-brain barrier to attenuate the inflammatory effects of HIV-1 infection in vitro. *J. Neuroimmune Pharmacol.* **2017**, *12*, 120–132. [CrossRef]

61. Ornatowski, W.; Lu, Q.; Yegambaram, M.; Garcia, A.E.; Zemskov, E.A.; Maltepe, E.; Fineman, J.R.; Wang, T.; Black, S.M. Complex interplay between autophagy and oxidative stress in the development of pulmonary disease. *Redox Biol.* **2020**, *36*, 101679. [CrossRef] [PubMed]

62. He, Z.; Zhang, Y.; Zhang, H.; Zhou, C.; Ma, Q.; Deng, P.; Lu, M.; Mou, Z.; Lin, M.; Yang, L.; et al. NAC antagonizes arsenic-induced neurotoxicity through TMEM179 by inhibiting oxidative stress in Oli-neu cells. *Ecotoxicol. Environ. Saf.* **2021**, *223*, 112554. [CrossRef]

63. Luo, Z.; Xu, X.; Sho, T.; Zhang, J.; Xu, W.; Yao, J.; Xu, J. ROS-induced autophagy regulates porcine trophectoderm cell apoptosis, proliferation, and differentiation. *Am. J. Physiol. Cell. Physiol.* **2019**, *316*, C198–C209. [CrossRef] [PubMed]

64. Liu, X.; Xu, L.; Shen, J.; Wang, J.; Ruan, W.; Yu, M.; Chen, J. Involvement of oxidative stress in tri-ortho-cresyl phosphate-induced autophagy of mouse Leydig TM3 cells in vitro. *Reprod. Biol. Endocrinol.* **2016**, *14*, 30. [CrossRef] [PubMed]

65. Li, Y.; Zhu, H.; Wang, S.; Qian, X.; Fan, J.; Wang, Z.; Song, P.; Zhang, X.; Lu, W.; Ju, D. Interplay of oxidative stress and autophagy in PAMAM dendrimers-induced neuronal cell death. *Theranostics* **2015**, *5*, 1363–1377. [CrossRef] [PubMed]

Article

PTPRN Serves as a Prognostic Biomarker and Correlated with Immune Infiltrates in Low Grade Glioma

Peng Li [1,†], Fanfan Chen [2,†], Chen Yao [2], Kezhou Zhu [1], Bei Zhang [3] and Zelong Zheng [4,*]

[1] VCU Massey Cancer Center, Department of Human and Molecular Genetics, Institute of Molecular Medicine, School of Medicine, Virginia Commonwealth University, Richmond, VA 23298, USA; peng.li@vcuhealth.org (P.L.); kezhou.zhu@vcuhealth.org (K.Z.)
[2] Neurosurgical Department, Shenzhen Second People's Hospital, The First Affiliated Hospital of Shenzhen University, Shenzhen 518035, China; cff_126com@126.com (F.C.); yaochen_1987@126.com (C.Y.)
[3] Department of Biostatistics, Virginia Commonwealth University, Richmond, VA 23298, USA; zhangb11@vcu.edu
[4] Department of Neurosurgery, Guangzhou First People's Hospital, School of Medicine, South China University of Technology, Guangzhou 510180, China
[*] Correspondence: zhengzelong-neurosurgery@hotmail.com
[†] These authors contributed equally.

Abstract: Background: Glioma is one of the most common malignant tumors of the central nervous system. Immune infiltration of tumor microenvironment was associated with overall survival in low grade glioma (LGG). However, effects of Tyrosine phosphatase receptor type N (PTPRN) on the progress of LGG and its correlation with tumor infiltration are unclear. Methods: Here, datasets of LGG were from The Cancer Genome Atlas (TCGA) and normal samples were from GTEx dataset. Gepia website and Human Protein Atlas (HPA) Database were used to analyze the mRNA and protein expression of PTPRN. We evaluated the influence of PTPRN on survival of LGG patients. MethSurv was used to explore the expression and prognostic patterns of single CpG methylation of PTPRN gene in LGG. The correlations between the clinical information and PTPRN expression were analyzed using logistic regression and Multivariate Cox regression. We also explored the correlation between PTPRN expression and cancer immune infiltration by TIMER. Gene set enrichment analysis (GSEA) was formed using TCGA RNA-seq datasets. Results: PTPRN mRNA and protein expression decreased in LGG compared to normal brain tissue in TCGA and HPA database. Kaplan-Meier analysis showed that the high expression level of PTPRN correlated with a good overall survival (OS) of patients with LGG. The Multivariate Cox analysis demonstrated that PTPRN expression and other clinical-pathological factors (age, WHO grade, IDH status, and primary therapy outcome) significantly correlated with OS of LGG patients. The DNA methylation pattern of PTPRN with significant prognostic value were confirmed, including cg00672332, cg06971096, cg01382864, cg03970036, cg10140638, cg16166796, cg03545227, and cg25569248. Interestingly, PTPRN expression level significantly negatively correlated with infiltrating level of B cell, CD4+ T cells, Macrophages, Neutrophils, and DCs in LGG. Finally, GSEA showed that signaling pathways, mainly associated with tumor microenvironment and immune cells, were significantly enriched in PTPRN high expression. Conclusion: PTPRN is a potential biomarker and correlates with tumor immune infiltration in LGG.

Keywords: PTPRN; glioma; immune infiltration; biomarker; tumor microenvironment

Citation: Li, P.; Chen, F.; Yao, C.; Zhu, K.; Zhang, B.; Zheng, Z. PTPRN Serves as a Prognostic Biomarker and Correlated with Immune Infiltrates in Low Grade Glioma. *Brain Sci.* **2022**, *12*, 763. https://doi.org/10.3390/brainsci12060763

Academic Editors: Hailiang Tang and Lucia Lisi

Received: 17 March 2022
Accepted: 24 May 2022
Published: 10 June 2022

Publisher's Note: MDPI stays neutral with regard to jurisdictional claims in published maps and institutional affiliations.

1. Introduction

Glioma is one of the most common malignant tumors of the central nervous system. According to the glioma grading classification criteria of the World Health Organization (WHO), gliomas are divided into Grade 1, 2, 3, and 4 [1,2]. Grade 1 and 2 gliomas belong to low-grade gliomas (LGGs) while Grade 3 and 4 gliomas are high-grade gliomas (HGGs). LGG encompasses a diverse group of diffusely infiltrative, slowly growing glial brain tumors that would dedifferentiate and progress to HGG. Standard treatment strategies for LGG include surgical treatment, radiotherapy, chemotherapy, and combination therapy. Despite advances in surgical techniques and the availability of new chemotherapeutic agents, outcomes for patients with LGG remain poor.

Cancer immunotherapy has recently become an important pillar of cancer treatment. Immunotherapy based on cytotoxic T lymphocyte-associated antigen 4 (CTLA4), programmed death-1 (PD-1) and programmed death ligand-1 (PD-L1) inhibitors have emerged as an effective treatment in melanoma, non-small cell lung carcinoma and glioma [3–5]. Tumor-infiltrating lymphocytes, such as tumor-associated macrophages (TAMs), play a very important role in patient prognosis and the efficacy of immunotherapy [6]. However, the discovery of glioma immunotherapy mainly affects GBM and rarely appears in LGG research. Additionally, in recent years, with the introduction of the concept of precision medicine, the detection of molecular diseases and the improvement of targeted therapy have significantly improved the overall survival rate of LGG patients. In patients with LGG, 1p/19q co-deletion, IDH mutation, and TERTp (telomerase reverse transcriptase gene promoter region) mutation can help assess the prognosis [7,8]. LGG patients with IDH1 mutation or 1p/19q co-deletion but not TERTp mutation can benefit from radiotherapy and chemotherapy. Although molecular targeted therapy has shown good clinical effects, the curing of LGG patients is still a challenge, especially due to the development of drug resistance. Therefore, there is an urgent need to develop a new and effective biomarker for the diagnosis and treatment of patients with LGG.

PTPRN, also called islet cell autoantigen 512 (ICA512/IA-2), is located on human chromosome band 2q35. PTPRN is mainly expressed in endocrine cells, neurons of the autonomic nervous system, and neuroendocrine neurons of the brain, including pancreas, pituitary, adrenal medulla, amygdala, and hypothalamus, because they all contain neurosecretory granules [9]. PTPRN is a type I transmembrane protein that participates in neuroendocrine processes, such as biogenesis, transport, and/or regulation of exocytosis [10]. It's involved in regulating the secretion pathways of various neuroendocrine cells and in the occurrence and development of diabetes mellitus [11]. Multiple studies have concluded that PTPRN may play a potential role in solid tumors [12–16]. PTPRN was identified as an independent prognostic factor in hepatocellular carcinoma [16]. Abnormal hypermethylation of PPTRN DNA is related to the OS of patients with ovarian cancer [12]. In addition, the expression of PTPRN mRNA and protein increased in breast cancer cells under hypoxia [13]. PTPRN was overexpressed in a group of glioblastoma patients, indicating poor survival [14]. Furthermore, PTPRN could be a biomarker that predicts the prognosis of glioblastoma and could estimate the OS of patients with glioblastoma [15]. However, the role of PTPRN in LGG is still unknown, especially its role in the regulation of tumor microenvironments.

In this study, the clinical features and survival information of patients with LGG from TCGA was analyzed using bioinformatics to assess the prognostic significance of PTPRN in LGG. We also investigate the relationship between PTPRN expression and tumor-infiltrating immune cells in LGG. Finally, the functional enrichment analysis was performed based on the differential genes expression in low and high PTPRN expression in LGG. These results shed light on the essential role of PTPRN and provide a potential mechanism for LGG.

2. Material and Methods

2.1. PTPRN Expression and Overall Survival Analysis by Gepia

Using an open website, Gepia (http://gepia2.cancer-pku.cn/#analysis, accessed on 9 May 2022), to analyze the mRNA expression of PTPRN between normal patient samples and LGG samples. All the samples are from TCGA ($n = 518$) and GTEx ($n = 207$) database. Fragments per kilobase million (FPKM) values were transformed into per kilobase million (TPM) values, which was more comparable between samples. A P value less than 0.05 was considered significant by Wilcoxon rank sum test. The samples of LGG were divided into two groups according to the median PTPRN expression to plot a Kaplan-Meier survival curve of LGG patients ($n = 257$ in both high and low expression groups, two cases were missing during follow-up).

2.2. The Human Protein Atlas Database

The human protein atlas (https://www.proteinatlas.org, accessed on 17 February 2022) is a Swedish-based program to map all the human proteins in cells, tissues, and organs using an integration of various omics technologies, including, antibody-based imaging, mass spectrometry-based proteomics, transcriptomics, and systems biology. It was used to compare the protein expression of PTPRN in LGG and normal brain tissue.

2.3. Clinic Data Analysis

We extracted the clinic data of the LGG project from TCGA (https://portal.gdc.cancer.gov/, accessed on 1 February 2022) [17]. The correlations between PTPRN expression and clinic-pathological parameters (histological type, IDH status, 1p/19q codeletion, and WHO grade) in LGG patients were analyzed [18].

2.4. DNA Methylation of the PTPRN Gene

DNA methylation plays an important function in prognostic evaluation and potential biomarker in tumorigenesis and progression. We used an integrated online tool, MethSurv (https://biit.cs.ut.ee/methsurv/, accessed on 2 May 2022), to explore the expression and prognostic patterns of single CpG methylation of PTPRN gene in LGG [19–21]. The DNA methylation results of PTPRN and survival analysis were generated via the MethSurv platform.

2.5. TIMER Database to Analyze Tumor Infiltrating Immune Cells

The TIMER database (https://cistrome.shinyapps.io/timer/, accessed on 5 February 2022), which includes 10,897 samples across 32 cancer types from TCGA, is a comprehensive resource for estimating the abundance of immune infiltrates cells, including B cells, CD4+ T cells, CD8+ T cells, neutrophils, macrophages, and DCs [22]. We analyzed the correlation of PTPRN expression with the abundance of immune infiltrates via the gene module. For each survival analysis, TIMER outputs Kaplan-Meier plots for TIICs and genes to visualize the survival differences between the upper and lower 50 percentile of patients. A log-rank p-value is also calculated and displayed for each Kaplan-Meier plot.

2.6. Gene Set Enrichment Analysis

A computational method that analyzes the statistical significance of a priori defined set of genes and the existence of concordant differences between two biological states is known as the GSEA [23]. In this study, GSEA created an initial list on the classification of the genes according to their correlation with the PTPRN expression. This computational method expounded on the noteworthy differences that we observed in the survival between high- and low- PTPRN expression groups. For each analysis, we performed 1000 repetitions of gene set permutations. The phenotype label that we put forth was the expression level of PTPRN. Additionally, to sort the enriched pathways in each phenotype, we utilized the nominal p-value and normalized enrichment score (NES). Gene sets with a discovery rate (FDR) <0.050 were considered to be significantly enriched.

2.7. Statistical Analysis

All p-values were two-sided, and values lower than 0.050 were considered significant. The correlations between the clinical information and PTPRN expression were analyzed using logistic regression. Multivariate Cox analysis was used to evaluate the influence of PTPRN expression and other clinic-pathological factors on survival. The CIBERSORT package was used to explore the difference in immune cells subtypes.

3. Results

3.1. PTPRN Expression Levels and Prognostic Value in LGG

We used Gepia website to determine the expression difference of PTPRN mRNA level between LGG and normal brain tissues. Compared with that in normal tissues, PTPRN mRNA expression was significantly decreased ($p < 0.050$) (Figure 1A) in LGG. Meanwhile, the protein of PTPRN expression was consistent with the mRNA results (Figure 1B). Kaplan-Meier analysis showed that high expression levels of PTPRN correlated with a good overall survival (OS) of patients with glioma ($p = 0.000$) (Figure 1C).

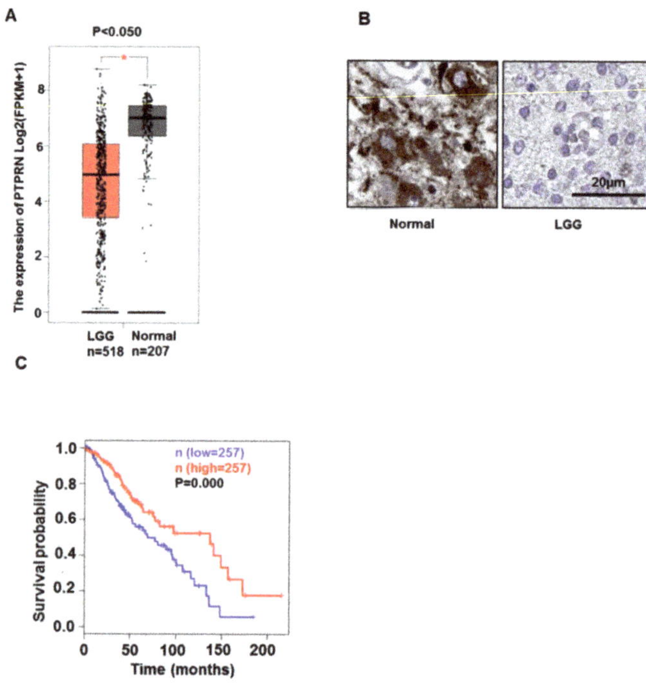

Figure 1. PTPRN mRNA and protein expression between normal and LGG samples, and overall survival in LGG. (**A**) PTPRN mRNA expression is higher in normal brain ($n = 207$) than that in LGG ($n = 518$) from the Gepia website (http://gepia2.cancer-pku.cn/#analysis, accessed on 9 May 2022). mean $\pm$ SD, *: $p < 0.050$. (**B**) Representative IHC images of PTPRN protein expression in normal brain tissues and glioma tissue (https://www.proteinatlas.org, accessed on 17 February 2022). (**C**) Overall patient survival in groups of high (red) and low (blue) expression was analyzed by Kaplan-Meier survival curve.

Then, we explored the associations between PTPRN expression and clinical characteristics, such as histological type, IDH status, 1p/19q status, and grade. In oligodendroglioma, PTPRN expression was sharply higher than astrocytoma ($p = 0.000$) (Figure 2A). Additionally, PTPRN expression was significantly higher in IDH mutation status ($p = 0.000$) (Figure 2B) and 1p/19q codeletion status gliomas ($p = 0.000$) (Figure 2C). However, the expression of PTPRN was similar in different grades of glioma ($p = 0.101$) (Figure 2D). Moreover, we examined whether PTPRN expression was an independent prognostic factor for LGG using Multivariate Cox regression analyses. The Multivariate Cox analysis demonstrated that age ($p = 0.000$), WHO grade ($p = 0.000$), IDH status ($p = 0.001$), primary therapy outcome ($p = 0.000$), and PTPRN expression ($p = 0.016$) significantly correlated with OS of LGG patients (Table 1). These results indicate that PTPRN expression is an independent prognostic index and high expression of PTPRN is correlated with longer OS.

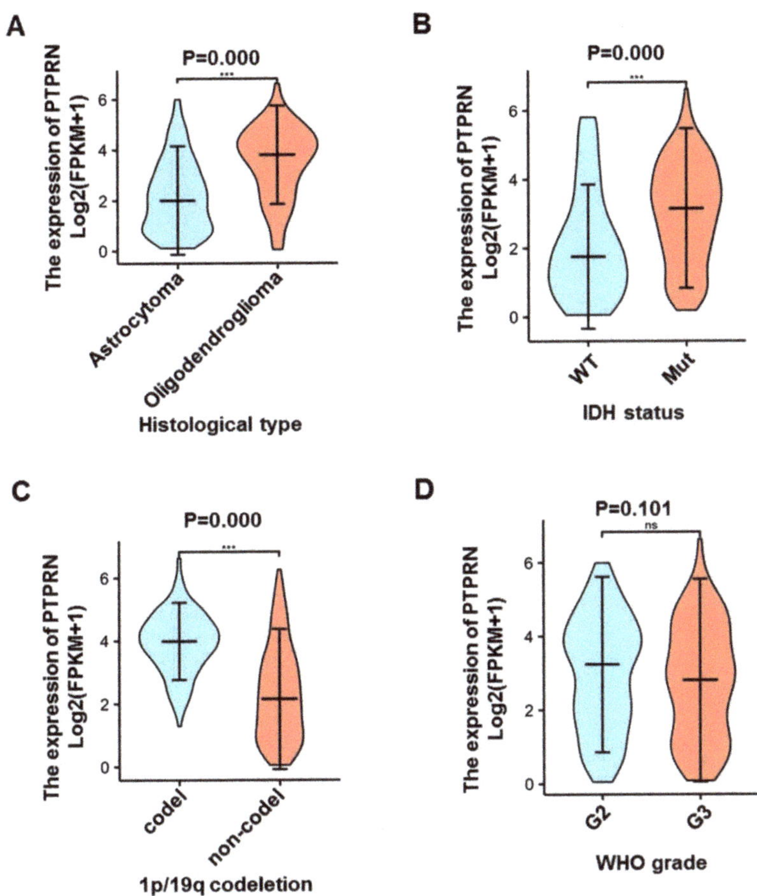

Figure 2. The associations between PTPRN mRNA expression and clinic characteristics. (**A**) In oligo-dendroglioma, PTPRN expression was higher than astrocytoma ($p = 0.000$). n (Astrocytoma) = 195, n (Oligodendroglioma) = 199. (**B**) PTPRN expression was high in IDH mutation status ($p = 0.000$) compare to IDH wildtype. n (WT) = 95, n (Mut) = 428. (**C**) PTPRN expression was elevated in 1p/19q codeletion status gliomas compare to Non-codel status ($p = 0.000$). n (Codel) = 171, n (Non-codel) = 357. (**D**) The expression of PTPRN was similar in different grades of glioma ($p = 0.101$). n (G2) = 224, n (G3) = 243. ns: not significant; ***, $p < 0.001$. abbreviation: WT: wildtype; Mut: mutation; Codel: codeletion; Non-codel: non-codeletion.

Table 1. Multivariate Cox regression analysis of clinicopathological features (including PTPRN expression) with OS in the TCGA datasets.

Characteristics	Total	HR(95% CI)		p Value
Age ($\leq$40 vs. >40)	527	3.245 (2.039–5.164)		<0.001
WHO grade (G2 vs. G3)	466	2.163 (1.355–3.450)		0.001
Histological type (Astrocytoma vs. Oligodendroglioma)	527	0.988 (0.624–1.564)		0.958
IDH status (WT vs. Mut)	524	0.391 (0.241–0.635)		<0.001
1p/19q codeletion (codel vs. non-codel)	527	1.731 (0.924–3.241)		0.087
Primary therapy outcome (PD&SD vs. PR&CR)	457	0.254 (0.134–0.481)		<0.001
PTPRN (High vs. Low)	527	0.872 (0.781–0.974)		0.016

3.2. DNA Mythylation Analysis of the PTPRN Gene in LGG

DNA methylation is an epigenetic alteration that relates to the tumorigenesis and progression. DNA methyltransferases on CpG island methylation are transcription factors that can suppress or promote cell growth and the process is a reversible. We showed the heatmap of DNA methylation clustering the expression levels of the PTPRN gene in LGG (Figure 3A). Furthermore, the DNA methylation pattern of PTPRN with significant prognostic value were also confirmed, such as cg00672332, cg06971096, cg01382864, cg03970036, cg10140638, cg16166796, cg03545227, and cg25569248 (Figure 3B–I, and Table 2).

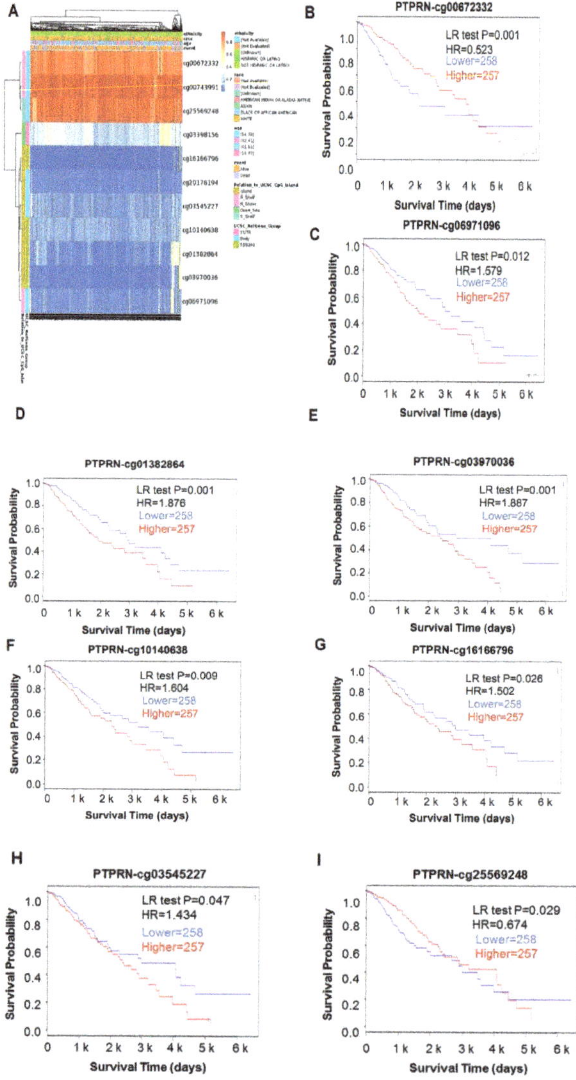

Figure 3. The DNA methylation of PTPRN in LGG of TCGA. (**A**) Heatmap of DNA methylation expression levels of the PTPRN gene in LGG by MethSurv platform. (**B–I**) Prognostic values of single CpG of the PTPRN gene in LGG. The threshold of significance was LR Test *p* value <0.05. cg00672332, cg06971096, cg01382864, cg03970036, cg10140638, cg16166796, cg03545227, and cg25569248 of PTPRN displays the significant level of DNA methylation in LGG.

Table 2. Prognostic Value of Single CpG of the PTPRN gene family in LGG by MethSurv platform.

Gene-CpG	HR	LR Test *p*-Value
PTPRN-Body-N_Shelf-cg00672332	0.523	**0.000** *
PTPRN-Body-N_Shore-cg00743991	0.817	0.260
PTPRN-Body-N_Shore-cg06971096	1.579	**0.012** *
PTPRN-TSS200-Island-cg01382864	1.876	**0.001** *
PTPRN-TSS200-Island-cg03970036	1.887	**0.001** *
PTPRN-TSS200-Island-cg10140638	1.604	**0.009** *
PTPRN-TSS200-Island-cg16166796	1.502	**0.026** *
PTPRN-Body-Island-cg03545227	1.434	**0.047** *
PTPRN-Body-Island-cg20176194	1.001	1.000
PTPRN-3'UTR-Open_Sea-cg03398156	0.873	0.450
PTPRN-Body-S_Shelf-cg25569248	0.674	**0.029** *

*: $p < 0.050$.

3.3. Correlation of PTPRN Expression with Immune Infiltration Level and Cumulative Survival in LGG

As mentioned above, some tumor-infiltrating lymphocytes are independent predictors of cancer survival, and thus, we investigated the association of PTPRN expression and immune infiltration levels in LGG. We implemented it by selecting PTPRN expression levels that were positively correlated with tumor purity. The investigation showed that the level of PTPRN expression negatively correlated with the infiltration level of B cell ($r = -0.370$, $p = 0.000$), CD4+ T cells ($r = -0.518$, $p = 0.000$), Macrophages ($r = -0.485$, $p = 0.000$), Neutrophils ($r = 0.324$, $p = 0.000$) and DCs ($r = -0.403$, $p = 0.000$) in LGG (Figure 4A). Moreover, our findings showed that B cell ($p = 0.000$), CD8+ T cells ($p = 0.010$), CD4+ T cells ($p = 0.000$), Macrophages ($p = 0.000$), Neutrophils ($p = 0.000$) and DCs ($p = 0.001$) are factors related to the cumulative survival rate of LGG over time (Figure 4B). These data strongly indicate that PTPRN is associated with immune infiltration in LGG.

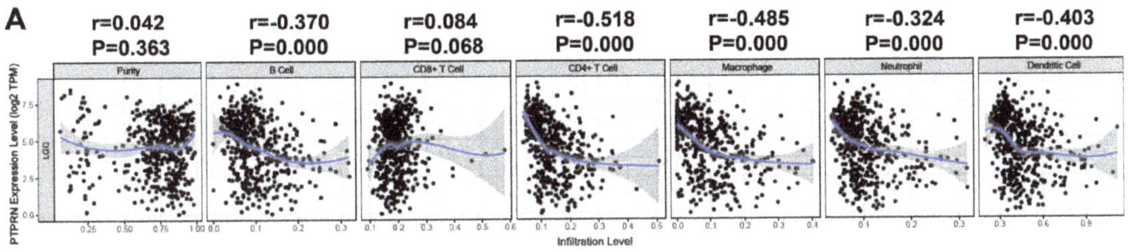

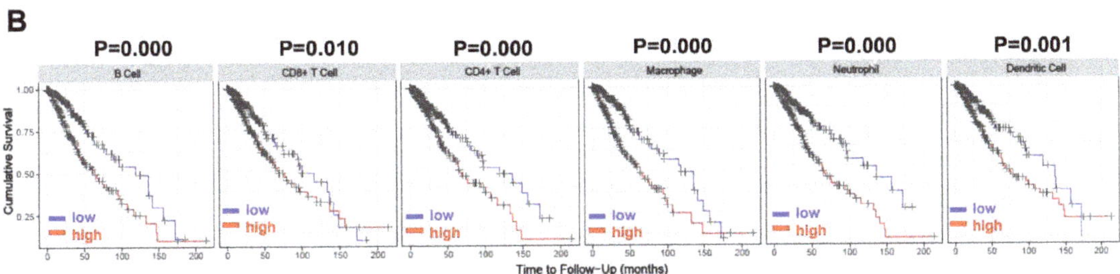

Figure 4. Correlation of PTPRN expression with immune infiltration level in LGG. (**A**) PTPRN expression level has significant negative correlations with infiltrating levels of B cell ($r = -0.370$, $p = 0.000$), CD4+ T cells ($r = -0.518$, $p = 0.000$), Macrophages ($r = -0.485$, $p = 0.000$), Neutrophils ($r = -0.324$, $p = 0.000$) and DCs ($r = -0.403$, $p = 0.001$) in LGG. (**B**) Cumulative survival is related to B cell ($p = 0.000$), CD8+ T cells ($p = 0.010$), CD4+ T cells ($p = 0.000$), Macrophages ($p = 0.000$), Neutrophils ($p = 0.000$) and DCs ($p = 0.001$) in LGG.

3.4. Gene Sets Enriched in PTPRN Expression Phenotype

Based on GSEA, PTPRN-associated signaling pathways were used to determine in LGG between low and high expression data sets and demonstrated significant differences (FDR < 0.050, p-value < 0.050) in the enrichment of GO and KEGG collection.

Ten pathways, related to the tumor microenvironment, including interferon signaling, immunoregulatory interactions between lymphoid and lymphoid cell, complement cascade, FCGR3A mediated IL10 synthesis, chemokine receptors bind chemokine, IL3, IL-5 and GM-CSF signaling, type II interferon signaling, Toll-like receptor signaling, NKT pathway, and TNFs bind their physiological receptors were showed significantly differential enrichment in PTPRN high and low expression groups based on NES, FDR, and p-value (Figure 5, Table 3). Strikingly, ten pathways, associated with immune cells, such as B cell receptor, TCR signaling, interactions between immune cells and microRNAs in the tumor microenvironment, T helper pathway, TCRA pathway, DC pathway, CTL pathway, B lymphocyte pathway, ASB cell pathway, and granulocytes pathway were demonstrated significantly differential enrichment between PTPRN high and low expression groups according to NES, FDR, and p-value (Figure 6, Table 4).

Table 3. Pathways, related to the microenvironment, were showed significantly differential enrichment in PTPRN high and low expression groups based on NES, FDR, and p-value.

ID	NES	p. Adjust	FDR
REACTOME_INTERFERON_SIGNALING	−2.084	0.048	0.038
REACTOME_IMMUNOREGULATORY_INTERACTIONS_BETWEEN_A_LYMPHOID_AND_A_NON_LYMPHOID_CELL	−2.835	0.045	0.036
REACTOME_COMPLEMENT_CASCADE	−2.976	0.041	0.033
REACTOME_FCGR3A_MEDIATED_IL10_SYNTHESIS	−2.496	0.04	0.032
REACTOME_CHEMOKINE_RECEPTORS_BIND_CHEMOKINES	−2.227	0.04	0.032
REACTOME_INTERLEUKIN_3_INTERLEUKIN_5_AND_GM_CSF_SIGNALING	−2.098	0.039	0.031
WP_TYPE_II_INTERFERON_SIGNALING_IFNG	−2.13	0.039	0.031
WP_TOLLLIKE_RECEPTOR_SIGNALING_RELATED_TO_MYD88	−1.902	0.039	0.031
BIOCARTA_NKT_PATHWAY	−2.118	0.039	0.031
REACTOME_TNFS_BIND_THEIR_PHYSIOLOGICAL_RECEPTORS	−1.931	0.039	0.031

significance: False discovery rate (FDR) < 0.050 and p. adjust < 0.050.

Table 4. Pathways, associated with immune cells, were demonstrated significantly differential enrichment between PTPRN high and low expression groups according to NES, FDR, and p-value.

ID	NES	p. Adjust	FDR
REACTOME_SIGNALING_BY_THE_B_CELL_RECEPTOR_BCR_	−2.039	0.045	0.036
REACTOME_TCR_SIGNALING	−2.041	0.044	0.035
WP_INTERACTIONS_BETWEEN_IMMUNE_CELLS_AND_MICRORNAS_IN_TUMOR_MICROENVIRONMENT	−2.264	0.039	0.031
BIOCARTA_THELPER_PATHWAY	−2.076	0.039	0.031
BIOCARTA_TCRA_PATHWAY	−2.271	0.039	0.031
BIOCARTA_DC_PATHWAY	−1.972	0.039	0.031
BIOCARTA_CTL_PATHWAY	−1.983	0.039	0.031
BIOCARTA_BLYMPHOCYTE_PATHWAY	−2.193	0.039	0.031
BIOCARTA_ASBCELL_PATHWAY	−2.062	0.039	0.031
BIOCARTA_GRANULOCYTES_PATHWAY	−2.162	0.039	0.031

significance: False discovery rate (FDR) < 0.050 and p. adjust < 0.050.

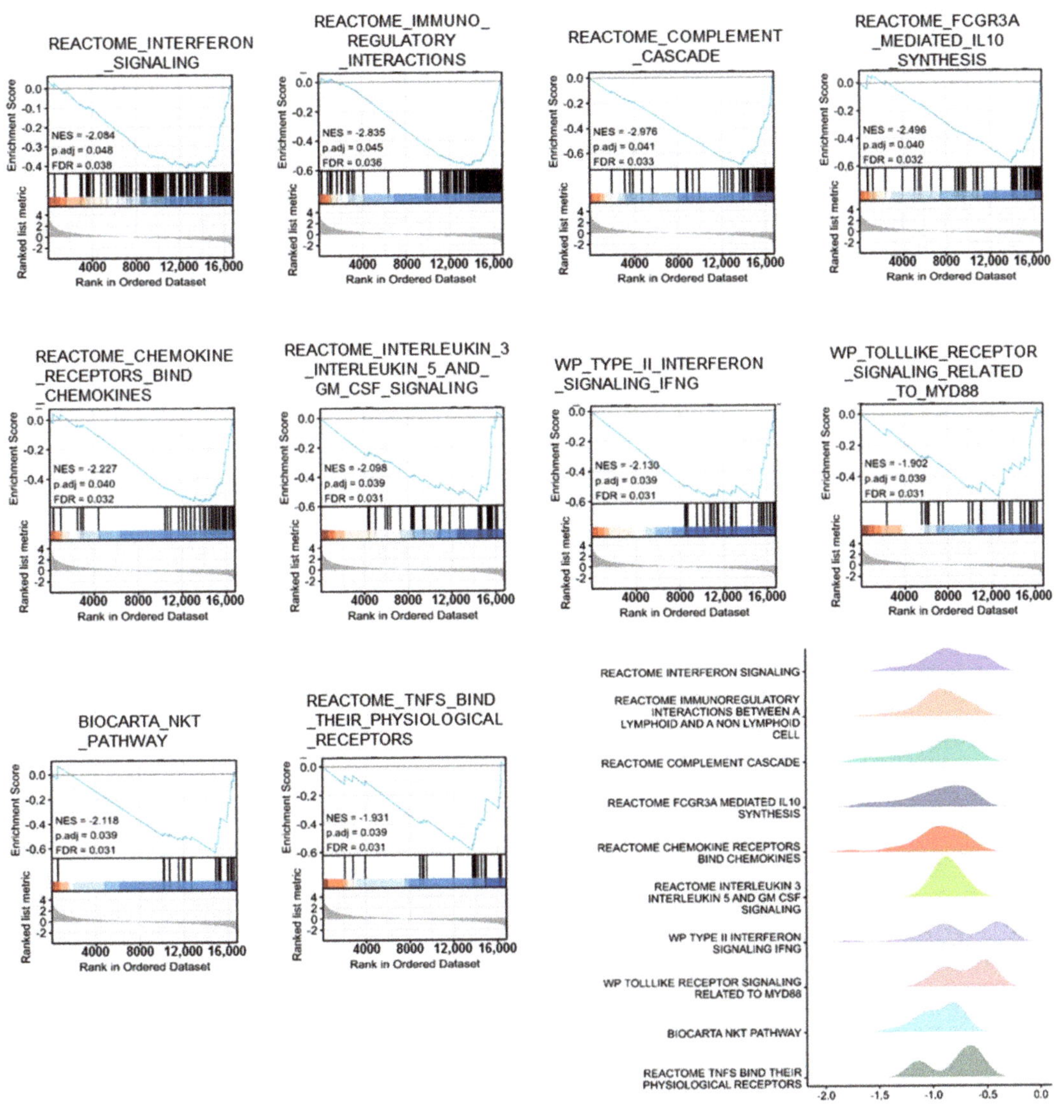

Figure 5. Pathways, related to the microenvironment, including interferon signaling, immunoregulatory interactions between lymphoid and lymphoid cell, complement cascade, FCGR3A mediated IL10 synthesis, chemokine receptors bind chemokine, IL3 IL-5 and GM-CSF signaling, type II interferon signaling, Toll-like receptor signaling, NKT pathway, and TNFs bind their physiological receptors were showed significantly differential enrichment in PTPRN high and low expression groups based on NES, FDR, and *p*-value.

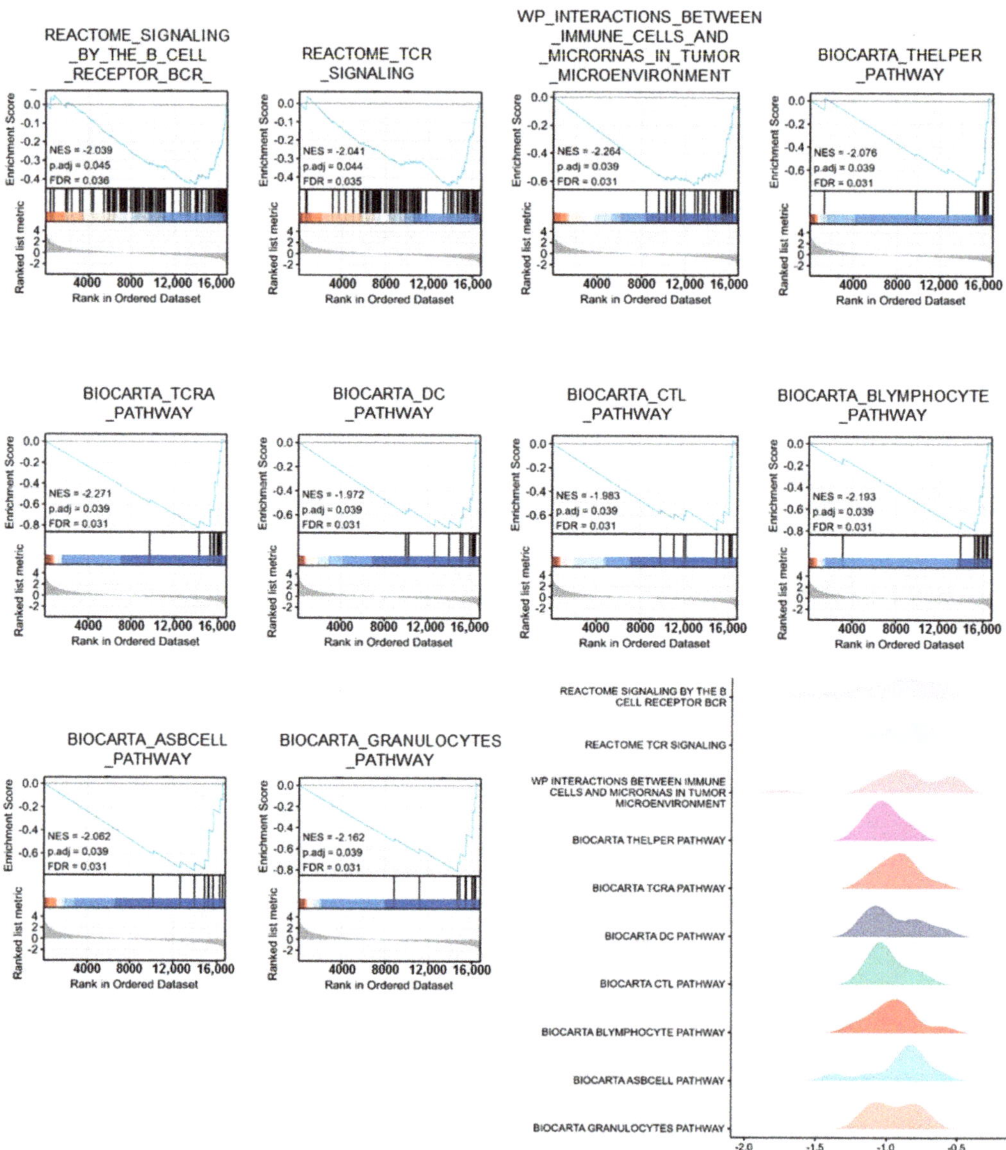

Figure 6. Pathways, associated with immune cells, such as B cell receptor, TCR signaling, interactions between immune cells and microRNAs in the tumor microenvironment, T helper pathway, TCRA pathway, DC pathway, CTL pathway, B lymphocyte pathway, ASB cell pathway, and granulocytes pathway were demonstrated significantly differential enrichment between PTPRN high and low expression groups according to NES, FDR, and *p*-value.

4. Discussion

PTPRN is a transmembrane protein and is expressed in different tumors, including breast cancer, hepatocellular carcinoma, ovarian cancer, and glioblastoma. However, little is known about the potential prognostic impact of PTPRN in LGG. In this investigation, we designed the in silico experiments to determine the clinical value of PTPRN in LGG.

Here, we used TCGA LGG data to analyze the expression level of PTPRN in both tumor and normal tissues. The result showed that mRNA and protein expression of PTPRN was sharply lower in LGG and LGG patients with high expression of PTPRN had a positive prognosis, which suggests LGG has prognostic value. Moreover, Multivariate COX regression analysis confirmed that high expression of PTPRN was an independent prognostic factor in patients with LGG. Previous studies have shown that the type of astrocytic tumor (vs. oligodendroglioma) is a poor prognostic parameter for LGG patients. In our study, we found that PTPRN expression was significantly lower in astrocytoma than in oligodendrocytoma, which again confirmed the prognostic value of PTPRN expression. Additionally, the relationship between PTPRN expression and IDH status or 1p/19q status was illustrated. The presence of IDH mutations is a powerful prognostic biomarker in patients with glioma and is associated with favorable outcomes independent of age and grade. Therefore, WHO strongly recommends that IDH phenotype is used as a new clinical diagnostic method. Complete deletion of both 1p and 19q (1p/19q co-deletion) is the molecular genetic signature of oligodendrogliomas. The presence of 1p/19q co-deletion is a strong independent prognostic biomarker associated with improved survival in both diffuse low-grade and anaplastic tumors. Our study demonstrated PTPRN expression was higher in LGG patients with IDH mutation and 1p/19q co-deletion. Additionally, DNA methylation plays an important role in prognostic evaluation and potential biomarker in tumorigenesis. the DNA methylation pattern of PTPRN with significant prognostic value were also confirmed, such as cg00672332, cg06971096, cg01382864, cg03970036, cg10140638, cg16166796, cg03545227, and cg25569248. In brief, these results imply that PTPRN could be a promising prognostic factor in LGG.

Immune infiltration in LGG is a hot topic at present. Understanding immune infiltrating cells is conducive to the development of immunotherapy for LGG. A significant association between PTPRN expression and immune infiltration levels in LGG was established. PTPRN expression level significantly negatively correlated with infiltrating level of B cell, CD4+ T cells, Macrophages, Neutrophils, and DCs in LGG. In addition, the relationship between PTPRN and immune cells implicates PTPRN plays a crucial part in regulating the immune microenvironment of LGG. To further investigate the function of PTPRN in LGG, we performed GSEA using TCGA dataset. GSEA showed that signaling pathways, mainly associated with microenvironment and immune cells, were significantly enriched in PTPRN high expression. Our overall finding emphasized the significant effect of PTPRN in immune infiltration in LGG.

PTPRN, termed ICA512 or ICA3 as well, was reported to play a role in vesicle-mediated secretory processes in hippocampus, pituitary, and pancreatic islets and regulate catalytic active protein-tyrosine phosphatases [24–26]. Interestingly, PTPRN is highly expressed in patients with high-grade glioma, which activates the PI3K/AKT pathway by interacting with HSP90AA1 to promote cell proliferation and metastasis [10]. However, a Kaplan-Meier survival analysis indicated that expression of PTPRN was downregulated in GBM tissue when compared with that of normal tissue and this gene was a good prognostic biomarker for GBM [27]. Although there is no detailed explanation for these controversial results except the sample numbers analyzed are small, we have reason to believe that PTPRN plays the important role in the development of pathophysiology in LGG.

There are several limitations in this study. First, the findings showed in this study need to be further verified by bench experiments. Second, the current findings only analyzed noticeable features associated with the prognosis of PTPRN without mechanism exploration. Therefore, additional research in vitro and in vivo is needed to confirm PTPRN efficacy as a viable target in the glioma immune microenvironment and to develop glioma immunotherapy in the future.

In summary, our findings demonstrated that high PTPRN expression correlates with good prognosis and restraints immune infiltration levels (in B cell, CD4+ T cell, macrophage, neutrophil, and dendritic cell) in LGG. In addition, there was a large degree of tumor immune cell infiltration in LGG patients with high PTPRN expression. In addition, the key pathways in LGG that were regulated by PTPRN are possibly tumor microenvironment and immune cells. These findings suggest that PTPRN could be an independent prognostic factor and correlates with tumor immune infiltration in LGG.

5. Conclusions

PTPRN could be an independent prognostic factor and correlates with tumor immune infiltration in LGG.

Author Contributions: P.L., F.C. and Z.Z. conceptualized and designed the study. P.L., B.Z. and K.Z. participated in the bioinformatics analysis. P.L. and C.Y. drafted the manuscript. All authors contributed to the article and approved the submitted version. All authors have read and agreed to the published version of the manuscript.

Funding: This work was supported by the Guangdong Basic and Applied Basic Research Foundation, under grant number 2019A1515011273.

Institutional Review Board Statement: Not applicable.

Informed Consent Statement: Not applicable.

Data Availability Statement: The data used to support the findings of this study are included with the article.

Conflicts of Interest: The authors have no conflicts of interest to declare.

Abbreviations

PTPRN: Tyrosine phosphatase receptor type N; LGG: low grade glioma; HPA: Human Protein Atlas; GSEA: Gene set enrichment analysis; OS: overall survival;CTLA4: cytotoxic T lymphocyte-associated antigen 4; PD-1: PD-L1: programmed death-1; programmed death ligand-1; TAMs: tumor-associated macrophages; TCGA: The Cancer Genome Atlas; aDC: activated DC; iDC: immature DC; pDC: Plasmacytoid DC; Tcm: T central memory; Tem: T effector memory; Tfh: T follicular helper; Tgd: T gamma delta; TIICs: tumor-infiltrating immune cells; THF: T follicular helper cells; TReg: Regulatory T cells; NES: normalized enrichment score.

References

1. Louis, D.N.; Perry, A.; Wesseling, P.A.-O.; Brat, D.A.-O.; Cree, I.A.; Figarella-Branger, D.; Hawkins, C.; Ng, H.K.; Pfister, S.M.; Reifenberger, G.; et al. The 2021 WHO Classification of Tumors of the Central Nervous System: A summary. *Neuro-Oncology* **2021**, *23*, 1231–1251. [CrossRef] [PubMed]

2. D'Amico, A.G.; Maugeri, G.; Vanella, L.; Pittalà, V.; Reglodi, D.; D'Agata, V. Multimodal Role of PACAP in Glioblastoma. *Brain Sci.* **2021**, *11*, 994. [CrossRef] [PubMed]

3. Werner, J.M.; Schweinsberg, V.; Schroeter, M.; von Reutern, B.; Malter, M.P.; Schlaak, M.; Fink, G.R.; Mauch, C.; Galldiks, N. Successful Treatment of Myasthenia Gravis Following PD-1/CTLA-4 Combination Checkpoint Blockade in a Patient with Metastatic Melanoma. *Front. Oncol.* **2019**, *9*, 84. [CrossRef]

4. Brahmer, J.R.; Govindan, R.; Anders, R.A.; Antonia, S.J.; Sagorsky, S.; Davies, M.J.; Dubinett, S.M.; Ferris, A.; Gandhi, L.; Garon, E.B.; et al. The Society for Immunotherapy of Cancer consensus statement on immunotherapy for the treatment of non-small cell lung cancer (NSCLC). *J. Immunother. Cancer* **2018**, *6*, 75. [CrossRef] [PubMed]

5. Litak, J.; Grajkowska, W.; Szumiło, J.; Krukow, P.; Maciejewski, R.; Roliński, J.; Grochowski, C. PD-L1 Expression Correlated with p53 Expression in Pediatric Glioblastoma Multiforme. *Brain Sci.* **2021**, *11*, 262. [CrossRef]

6. Bingle, L.; Brown, N.J.; Lewis, C.E. The role of tumour-associated macrophages in tumour progression: Implications for new anticancer therapies. *J. Pathol.* **2002**, *196*, 254–265. [CrossRef]

7. Eckel-Passow, J.E.; Lachance, D.H.; Molinaro, A.M.; Walsh, K.M.; Decker, P.A.; Sicotte, H.; Pekmezci, M.; Rice, T.; Kosel, M.L.; Smirnov, I.V. Glioma groups based on 1p/19q, IDH, and TERT promoter mutations in tumors. *N. Engl. J. Med.* **2015**, *372*, 2499–2508. [CrossRef]

8. Specchia, F.M.C.; Monticelli, M.; Zeppa, P.; Bianconi, A.; Zenga, F.; Altieri, R.; Pugliese, B.; Di Perna, G.; Cofano, F.; Tartara, F.; et al. Let Me See: Correlation between 5-ALA Fluorescence and Molecular Pathways in Glioblastoma: A Single Center Experience. *Brain Sci.* **2021**, *11*, 795. [CrossRef]

9. Torkko, J.M.; Primo, M.E.; Dirkx, R.; Friedrich, A.; Viehrig, A.; Vergari, E.; Borgonovo, B.; Sönmez, A.; Wegbrod, C.; Lachnit, M.; et al. Stability of proICA512/IA-2 and Its Targeting to Insulin Secretory Granules Require β4-Sheet-Mediated Dimerization of Its Ectodomain in the Endoplasmic Reticulum. *Mol. Cell. Biol.* **2015**, *35*, 914–927. [CrossRef]

10. Wang, D.; Tang, F.; Liu, X.; Fan, Y.S.; Zheng, Y.; Zhuang, H.; Chen, B.D.; Zhuo, J.; Wang, B. Expression and Tumor-Promoting Effect of Tyrosine Phosphatase Receptor Type N (PTPRN) in Human Glioma. *Front. Oncol.* **2021**, *11*, 676287. [CrossRef]

11. Fu, Z.; Gilbert, E.R.; Liu, D.M. Regulation of Insulin Synthesis and Secretion and Pancreatic Beta-Cell Dysfunction in Diabetes. *Curr. Diabetes Rev.* **2013**, *9*, 25–53. [CrossRef] [PubMed]

12. Bauerschlag, D.O.; Ammerpohl, O.; Brautigam, K.; Schem, C.; Lin, Q.; Weigel, M.T.; Hilpert, F.; Arnold, N.; Maass, N.; Meinhold-Heerlein, I.; et al. Progression-Free Survival in Ovarian Cancer Is Reflected in Epigenetic DNA Methylation Profiles. *Oncology* **2011**, *80*, 12–20. [CrossRef] [PubMed]

13. Liang, H.; Xiao, J.; Zhou, Z.; Wu, J.; Ge, F.; Li, Z.; Zhang, H.; Sun, J.; Li, F.; Liu, R.; et al. Hypoxia induces miR-153 through the IRE1α-XBP1 pathway to fine tune the HIF1α/VEGFA axis in breast cancer angiogenesis. *Oncogene* **2018**, *37*, 1961–1975. [CrossRef] [PubMed]

14. Shergalis, A.; Bankhead, A.; Luesakul, U.; Muangsin, N.; Neamati, N. Current Challenges and Opportunities in Treating Glioblastoma. *Pharmacol. Rev.* **2018**, *70*, 412–445. [CrossRef]

15. Yin, W.; Tang, G.H.; Zhou, Q.W.; Cao, Y.D.; Li, H.X.; Fu, X.Y.; Wu, Z.P.; Jiang, X.J. Expression Profile Analysis Identifies a Novel Five-Gene Signature to Improve Prognosis Prediction of Glioblastoma. *Front. Genet.* **2019**, *10*, 419. [CrossRef]

16. Zhangyuan, G.Y.; Yin, Y.; Zhang, W.J.; Yu, W.W.; Jin, K.P.; Wang, F.; Huang, R.Y.; Shen, H.Y.; Wang, X.C.; Sun, B.C. Prognostic Value of Phosphotyrosine Phosphatases in Hepatocellular Carcinoma. *Cell. Physiol. Biochem.* **2018**, *46*, 2335–2346. [CrossRef]

17. Liu, J.; Lichtenberg, T.; Hoadley, K.A.; Poisson, L.M.; Lazar, A.J.; Cherniack, A.D.; Kovatich, A.J.; Benz, C.C.; Levine, D.A.; Lee, A.V.; et al. An Integrated TCGA Pan-Cancer Clinical Data Resource to Drive High-Quality Survival Outcome Analytics. *Cell* **2018**, *173*, 400–416. [CrossRef]

18. Ceccarelli, M.; Barthel, F.P.; Malta, T.M.; Sabedot, T.S.; Salama, S.R.; Murray, B.A.; Morozova, O.; Newton, Y.; Radenbaugh, A.; Pagnotta, S.M.; et al. Molecular Profiling Reveals Biologically Discrete Subsets and Pathways of Progression in Diffuse Glioma. *Cell* **2016**, *164*, 550–563. [CrossRef]

19. Anuraga, G.; Wang, W.-J.; Phan, N.N.; An Ton, N.T.; Ta, H.D.K.; Berenice Prayugo, F.; Minh Xuan, D.T.; Ku, S.-C.; Wu, Y.-F.; Andriani, V.; et al. Potential Prognostic Biomarkers of NIMA (Never in Mitosis, Gene A)-Related Kinase (NEK) Family Members in Breast Cancer. *J. Pers. Med.* **2021**, *11*, 1089. [CrossRef]

20. Modhukur, V.; Iljasenko, T.; Metsalu, T.; Lokk, K.; Laisk-Podar, T.; Vilo, J. MethSurv: A web tool to perform multivariable survival analysis using DNA methylation data. *Epigenomics* **2018**, *10*, 277–288. [CrossRef]

21. Zhang, Y.J.; Yu, C.R. Prognostic characterization of OAS1/OAS2/OAS3/OASL in breast cancer. *BMC Cancer* **2020**, *20*, 575. [CrossRef] [PubMed]

22. Li, T.W.; Fan, J.Y.; Wang, B.B.; Traugh, N.; Chen, Q.M.; Liu, J.S.; Li, B.; Liu, X.S. TIMER: A Web Server for Comprehensive Analysis of Tumor-Infiltrating Immune Cells. *Cancer Res.* **2017**, *77*, E108–E110. [CrossRef] [PubMed]

23. Subramanian, A.; Kuehn, H.; Gould, J.; Tamayo, P.; Mesirov, J.P. GSEA-P: A desktop application for Gene Set Enrichment Analysis. *Bioinformatics* **2007**, *23*, 3251–3253. [CrossRef] [PubMed]

24. Harashima, S.-I.; Horiuchi, T.; Wang, Y.; Notkins, A.L.; Seino, Y.; Inagaki, N. Sorting nexin 19 regulates the number of dense core vesicles in pancreatic β-cells. *J. Diabetes Investig.* **2012**, *3*, 52–61. [CrossRef]

25. Kawasaki, E.; Eisenbarth, G.S.; Wasmeier, C.; Hutton, J.C. Autoantibodies to protein tyrosine phosphatase-like proteins in type I diabetes—Overlapping specificities to phogrin and ICA512/IA-2. *Diabetes* **1996**, *45*, 1344–1349. [CrossRef]

26. Hermel, J.-M.; Dirkx, R.; Solimena, M. Post-translational modifications of ICA512, a receptor tyrosine phosphatase-like protein of secretory granules. *Eur. J. Neurosci.* **1999**, *11*, 2609–2620. [CrossRef]

27. Zhu, X.P.; Pan, S.; Li, R.; Chen, Z.B.; Xie, X.Y.; Han, D.Q.; Lv, S.Q.; Huang, Y.K. Novel Biomarker Genes for Prognosis of Survival and Treatment of Glioma. *Front. Oncol.* **2021**, *11*, 49. [CrossRef]

Article

TRP Family Genes Are Differently Expressed and Correlated with Immune Response in Glioma

Chaoyou Fang [1], Houshi Xu [1], Yibo Liu [2], Chenkai Huang [3], Xiaoyu Wang [2], Zeyu Zhang [2], Yuanzhi Xu [4], Ling Yuan [1], Anke Zhang [2], Anwen Shao [2] and Meiqing Lou [1,*]

[1] Department of Neurosurgery, Shanghai General Hospital, School of Medicine, Shanghai Jiao Tong University, Shanghai 200080, China; chaoyouf@163.com (C.F.); houshi@sjtu.edu.cn (H.X.); yuanling093@163.com (L.Y.)
[2] Department of Neurosurgery, Second Affiliated Hospital, School of Medicine, Zhejiang University, Hangzhou 310009, China; 17853735042@163.com (Y.L.); 21818274@zju.edu.cn (X.W.); 12018507@zju.edu.cn (Z.Z.); theanke@163.com (A.Z.); shaoanwen@zju.edu.cn (A.S.)
[3] Department of Molecular and Cellular Biology, University of California, Berkeley, CA 94720, USA; aly1573592486@aliyun.com
[4] Department of Neurosurgery, Huashan Hospital, Shanghai Medical College, Fudan University, Shanghai 200040, China; dr.yuanzhi.xu@gmail.com
* Correspondence: meiqing_lou2020@163.com; Tel.: +86-21-63240090; Fax: +86-21-63079925

Abstract: (1) Background: glioma is the most prevalent primary tumor of the human central nervous system and accompanies extremely poor prognosis in patients. The transient receptor potential (TRP) channels family consists of six different families, which are closely associated with cancer cell proliferation, differentiation, migration, and invasion. TRP family genes play an essential role in the development of tumors. Nevertheless, the function of these genes in gliomas is not fully understood. (2) Methods: we analyze the gene expression data of 28 TRP family genes in glioma patients through bioinformatic analysis. (3) Results: the study showed the aberrations of TRP family genes were correlated to prognosis in glioma. Then, we set enrichment analysis and selected 10 hub genes that may play an important role in glioma. Meanwhile, the expression of 10 hub genes was further established according to different grades, survival time, IDH mutation status, and 1p/19q codeletion status. We found that TRPC1, TRPC3, TRPC4, TRPC5, TRPC6, MCOLN1, MCOLN2, and MCOLN3 were significantly correlated to the prognosis in glioma patients. Furthermore, we illustrated that the expression of hub genes was associated with immune activation and immunoregulators (immunoinhibitors, immunostimulators, and MHC molecules) in glioma. (4) Conclusions: we proved that TRP family genes are promising immunotherapeutic targets and potential clinical biomarkers in patients with glioma.

Keywords: TRP family genes; glioma; prognosis; immunotherapy; bioinformatics analysis

Citation: Fang, C.; Xu, H.; Liu, Y.; Huang, C.; Wang, X.; Zhang, Z.; Xu, Y.; Yuan, L.; Zhang, A.; Shao, A.; et al. TRP Family Genes Are Differently Expressed and Correlated with Immune Response in Glioma. *Brain Sci.* 2022, 12, 662. https://doi.org/10.3390/brainsci12050662

Academic Editor: Hailiang Tang

Received: 15 April 2022
Accepted: 17 May 2022
Published: 19 May 2022

1. Introduction

As the most prevalent subgroup of primary intracranial tumors, glioma accounts for 24.5% of all primary brain tumors and 80.9% of malignant primary brain tumors [1,2]. According to the WHO Classification of CNS tumors, the diagnostic categories can partly be defined by genotype, and this was the first study to define glioma according to the presence or absence of IDH mutation and 1p/19q codeletion [3,4]. Currently, the treatment for glioma holds three predominant tenants: surgical resection, external beam radiation therapy, and chemotherapy. However, even for the most promising treatment, the effect of immunotherapy still falls far short of our expectations. Systemic medications cannot reach a therapeutic concentration inside solid tumor tissues and are accompanied by significant systemic side effects [5,6].

Currently, targeted therapy has become an important method in individualized treatment for glioma. Previous studies revealed that gliomas have a high immune infiltration rate [7,8]. Immune checkpoints are costimulators or cosuppressors required to produce an immune response [9]. Affecting immune checkpoints has achieved a breakthrough in the treatment of glioma. For instance, immune-checkpoint-blocking therapy is effective in a GBM mouse model, which suggests that immune checkpoints may be a potential clinical strategy for glioma [10]. Transient receptor potential (TRP) channels are widely expressed on the plasma membrane in numerous types of cells. It consists of six different families, including TRPC (TRP canonical), TRPV (TRP vanilloid), TRPM (TRP melastatin), TRPML (TRP mucolipin), TRPP (TRP polycystin), and TRPA (TRP ankyrin) [11]. TRP channels are cation channels, which are associated with cancer [12]. The TRP family genes related to various diseases and their role in cancer are of particular concern, such as breast cancer, prostate cancer, melanoma, and tract cancers [12–14]. mRNA levels for PARP1, PARP2, and TRPM2 genes are deregulated in acute myelogenous leukemia cells [15]. The plasma membrane of triple-negative breast cancer (TNBC) Cells MDA-MB-231 were found to overexpress TRPC3 [16]. Recently, the functions of TRP family genes in glioma have been of great importance. Evidence has shown that TRP family genes are associated with tumor proliferation, the cell cycle, apoptosis, resistance to treatment, and migration [17]. In glioma, TRP family genes could play either an oncogenic role or suppressive role in tumors. In a previous study, TRP family genes were demonstrated to play a crucial role in tumor death and therapy in glioma. For instance, TPRC3 suppresses cell proliferation and promotes cell death in glioma. A previous study revealed that TRPV2 can decrease cell proliferation in hGBM cells and GSCs and induce apoptosis and autophagy in tumor cells [18]. TRPML2 also plays a crucial role in the progression of glioma [17]. Although TRP family genes have an effect on tumorigenesis in a variety of tumors, the underlying mechanism of TRP family genes in glioma is still not very clear. TRPC1 channels may affect glioma cell division mainly by regulating calcium signaling during cytokinesis [19]. TRPC6 induces a sustained elevation of intracellular calcium in conjunction with activation of the calcineurin-nuclear factor of the activated T-cells (NFAT) pathway. By inhibiting the calcineurin-NFAT pathway, the development of malignant GBM phenotypes under hypoxia is significantly reduced [20].

The relationship between TRP family genes and inflammatory markers and inhibitory molecules have been proved in several studies. A fatty acid amide, palmitoylethanolamide (PEA), has been proven to have anti-inflammatory and analgesic properties by activating TRPV1 [21,22]. Bang et al. found that dimethylallyl pyrophosphate (DMAPP) induces acute inflammation by activating TRPV4 [23]. According to Zhao et al., Cathepsin S (Cat-S), which is released by antigen-presenting cells during inflammation, activates PAR2 and TRPV4 to cause inflammation [24]. Inflammation has been well established as a significant factor in cancer development [25]. The use of inflammation markers in cancer immunotherapy has been proven promising [26].

In our present study, we used data from The Tumor Genome Atlas (TCGA) database to analyze the expression of TRP family genes in glioma and their relationship with clinico-pathological characteristics. Moreover, we also performed bioinformatic analysis to explore the potential functions of the TRP family genes. The analysis showed that TRP family genes may be potential immunotherapeutic targets and novel clinical biomarkers for patients with glioma.

2. Materials and Methods

2.1. Genetic Alteration of TRP Family Genes in Glioma

CBioPortal (http://www.cbioportal.org/) (accessed on 7 November 2021) is a free web resource for visualizing and analyzing cancer genomics data in a multidimensional manner and was used to determine the correlation between aberrations of TRP family genes and survival time in glioma [27]. Additionally, the correlation was measured by KM diagrams. We selected sixteen cancers (including breast cancer, colon cancer, cholangiocarcinoma, glioma, kidney cancer, liver cancer, uterus corpus endometrial carcinoma, esophageal cancer, lung cancer, pancreatic cancer, stomach cancer, prostate cancer, head and neck cancer, thyroid cancer, skin cancer, and bladder cancer) that are more common and important in human and more abundant in the database. We included all relevant tumor samples in the database for analysis, including IDH1 wild-type and mutated tumors, 1p/19q codeleted tumors and those without this alteration, and also all WHO grades.

2.2. Enrichment Analysis and Hub Genes

To detect which sites and pathways the TRP family genes may act on, an enrichment analysis was conducted using GO and KEGG. By using the Bioinformatics database, GO and KEGG enrichment results were visualized using bubble diagrams. Diagrams of GO enrichment analysis included three parts—a biological process (BP) section, a cellular component (CC) section, and a molecular function (MF) section [28]. KEGG, a utility database resource, was used to understand advanced functions and biological systems [29]. In both analyses, GO enrichment and KEGG enrichment were calculated using $p < 0.05$ as the level of significance.

The hub genes were excavated by the STRING database (https://cn.string-db.org/) (accessed on 11 November 2021) and made by CytoHubba plug-in in Cytoscape software. By inputting the 28 TRP family genes in the STRING database, we obtained the data that can be made into a final figure in Cytoscape software. Interaction scores (median confidence) of 0.4 were considered cut-off values. The Cytoscape software was used to establish a network scaffold. Additionally, the CytoHubba plug-in in Cytoscape software can detect and locate the top 10 of the relevant TRP family genes by using the maximal clique centrality (MCC) and mark them in red (high correlation) or yellow (low correlation) colors based on their correlation with glioma.

2.3. Validation of Hub Genes

To validate the expression level of hub genes in different grades of HGG vs. LGG, IDH mutation status vs. IDH wildtype status, and 1p/19q codeletion status vs. 1p/19q non-deletion status, CGGA database (http://www.cgga.org.cn/) (accessed on 15 November 2021) was used for analysis [30]. Human Protein Atlas (HPA) database (The Human Protein Atlas), which contains immunohistochemistry-based expression data for over 20 types of cancers, was used to identify the protein expression levels of TRPC1, TRPC6, TRPM7, MCOLN1, MCOLN2, and MCOLN3 in HGG and LGG tissues [31].

2.4. Survival Analysis of Hub Genes

The correlation between overall survival (OS) time and hub genes expressed, in addition to disease-free survival (DFS) time and hub genes expression was assessed by the Kaplan–Meier (KM) survival analysis in the GEPIA database (http://gepia.cancer-pku.cn/) (accessed on 18 November 2021) in glioma patients. Additionally, the standard for gene expression analysis was the lower and upper 50% of gene expression, in which standard patients of glioma were categorized into 2 groups. Log-rank was regarded as significant when $p < 0.05$.

2.5. Immune Response Prediction

By retrieving literature, we summarized immune cells, immune cell markers, and immune checkpoint molecules, which might be related to glioma [32–34]. Using the TIMER2.0 database (http://timer.cistrome.org/) (accessed on 21 November 2021) [35], we detected the relationship between hub genes and the expression levels of immune cell infiltration (including purity, B cell, CD8+ T Cell, CD4+ T Cell, macrophage, neutrophil, dendritic cell), biomarkers (EGFR, EGFL7, CIC, BRAF, H3F3A, HIST1H3B, ATRX, DAXX, PDGFRA, TERT, CHI3L1, MYC, PTEN, S100A8, S100A9, AHSG, SOCS3, NPM1, FTL, BIRC5, GFAP, CXCR4, CTSD, LDHA, TP53, IL13RA2, PROM1, CD44), immune cell markers (CD8A, CD8B, CD3D, CD79A, CD86, CSF1R, CCL2, CD68, IL10, NOS2, IRF5, PTGS2, CD163, VSIG4, MS4A4A, CEACAM8, KIR2DL3, KIR2DL4, KIR3DL1, KIR3DL2, CD1C, NRP1, STAT1, IFNG, STAT3, IL17A, FOXP3, CCR8, STAT5B, TGFB1, PDCD1, GZMB, CD3E, CD2, CD19,

KIR3DL3, KIR2DS4, HLA-DPB1, ITGAM, CCR7, KIR2DL1, STAT6, STAT5A, IL13BCL6, ITGAX, TBX21, STAT4, CTLA4, LAG3, HAVCR2, HLA-DQB1, HLA-DRA, HLA-DPA1, TNF, GATA3, IL21) and immune checkpoint molecules (IDO1, LAG3, CD80, PDCD1LG2, CD86, PDCD1, LAIR1, IDO2, CD276, CD40, CD274, HAVCR2, CD28, CD48, VTCN1, CD160, CD44, TNFSF18, CD200R1, TNFSF4, CD200, NRP1, CTLA4, TNFRSF9, COS, TNFRSF8, TNFSF15, TNFRSF14, CD27,BTLA, LGALS9, TMIGD2, TNFRSF25, CD40LG, ADORA2A, TNFRSF4, TNFSF14, HHLA2, CD244, TNFRSF18, BTNL2, C10ORF54, TIGIT, CD70, TNFSF9, ICOSLG, KIR3DL1) in glioma. Subsequently, we selected the module of Gene_Corr in TIMER2.0, inputted the target genes, and then downloaded the correlated data. Using the TBtools application, after we selected the heatmap module from graphics, inputted the selected data, and modified some parameters, the final figures were obtained.

The correlations between the expression level of hub genes and immunoregulators, which contains immunoinhibitors, inmunostimulators, and MHC molecules, were calculated by the TISIDB database (http://cis.hku.hk/TISIDB/) (accessed on 22 November 2021) [36].

2.6. Statistical Data

p-Values of less than 0.05 wcrc considered statistically significant in all the analyses, except for those specifically mentioned.

3. Results

3.1. Aberrations in TRP Family Genes Are Correlated with Prognosis in Human Cancers

Using the cBioPortal database, we analyzed the correlation between TRP family aberrations and overall survival (OS) time to evaluate the functional importance of the TRP family in human cancers. As presented in Figure 1, aberrations in the TRP family in the genome were remarkably correlated with shorter OS time in patients with prostate cancer, lung cancer, and skin cancer; however, they were remarkably correlated with better prognosis in patients with cholangiocarcinoma, glioma, bladder cancer, and liver cancer. A significant correlation between the TRP family and survival time was not found in the remaining human cancers. As seen in Figure 1, we evaluated the prognostic usefulness of TRP family genes by pan-cancer analysis. Additionally, we found that aberrations of TRP family genes were associated with a better prognosis among glioma patients.

3.2. GO and KEGG Enrichment Analysis and Hub Genes of the TRP Family

To understand downstream pathways of TRP family genes in gliomas, GO and KEGG enrichment analyses were conducted using bioinformatics databases to analyze the potential biological functions of TRP family genes. As shown in Figure 2A, the most significant GO-enriched terms were involved in calcium ion transmembrane transport, cellular divalent inorganic cation homeostasis, and protein tetramerization (BP); cation channel complex, calcium channel complex, and cell body (CC); calcium channel activity, ligand-gated calcium channel activity, and store-operated calcium channel activity (MF). KEGG-enriched terms were mainly involved in the process of inflammatory mediator regulation of TRP channels, axon guidance, and mineral absorption (Figure 2A).

After the potential interaction among these genes was investigated, the CytoHubba plug-in in Cytoscape software was used to reveal the relationship among hub genes. The top 10 hub genes identified included TRPC1, MCOLN1, MCOLN2, MCOLN3, TRPC5, TRPC4, TRPA1, TRPC3, TRPC6, and TRPM7 (Figure 2B). By using the TIMER 2.0 database and TBtools software, the correlation between hub genes and tumor malignancy-associated genes in glioma was determined. As shown in Figure 2C, a majority of hub genes were positively correlated with tumor malignancy.

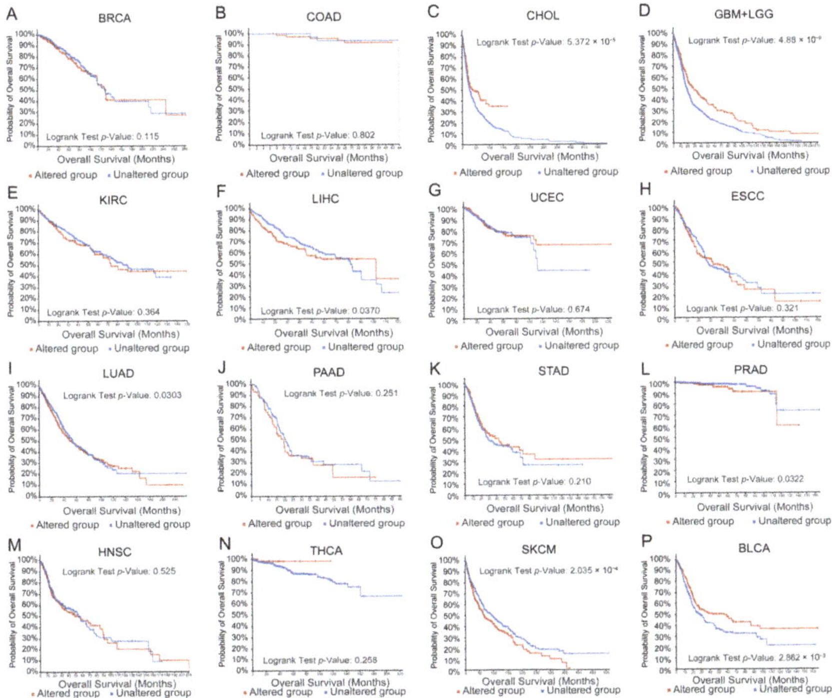

Figure 1. Human cancer prognosis was related to aberrations of genes of the TRP family genes. (**A–P**) The relationship between TRPs family genes aberrations and overall survival time in patients with breast cancer, colon cancer, cholangiocarcinoma, glioma, kidney cancer, liver cancer, uterus corpus endometrial carcinoma, esophageal cancer, lung cancer, pancreatic cancer, stomach cancer, prostate cancer, head and neck cancer, thyroid cancer, skin cancer, and bladder cancer.

3.3. Amplification, Mutation, and Deletion of Hub Genes in Glioma

The different functions of hub genes have been involved in several cancers in previous studies, such as pancreatic cancer [37], bladder cancer [38], esophageal squamous cell carcinoma [39], and non-small cell lung carcinoma [40], especially in glioma [41]. The purpose of this study was to explore the functions of TRP family genes in gliomas. CBioPortal was used to identify genetic variations in hub genes in 6853 cases. Among the 10 hub genes, different levels of genetic change were found, including TRPC1, MCOLN1, MCOLN2, MCOLN3, TRPC5, TRPC4, TRPA1, TRPC3, TRPC6, and TRPM7 (Figure 3). The results revealed that all hub genes were amplified, deleted, and mutated in glioma, among which MCOLN1 displayed the highest incidence rate (1.8%).

3.4. Expression Difference of Hub Genes in Different Grades of Glioma

The results from the GCCA database revealed that the gene expression levels of TRPC1, MCOLN1, MCOLN2, MCOLN3, TRPC5, TRPC4, TRPC3, TRPC6, and TRPM7 in different grades were significantly different. The gene expression levels of TRPC1, TRPC3, and TRPC5 were significantly higher in low-grade glioma (LGG), while the levels of TRPC4, TRPC6, TRPM7, MCOLN1, MCOLN2, and MCOLN3 were significantly higher in high-grade glioma (HGG). All of these results showed that high expression levels of TRPC4, TRPC6, TRPM7, MCOLN1, MCOLN2, and MCOLN3 may correlate with a worse prognosis, while higher expression of TRPC1, TRPC3, and TRPC5 correlates with a better prognosis. However, the expression of TRPC4 in glioma presented no statistical significance (Figure 4A).

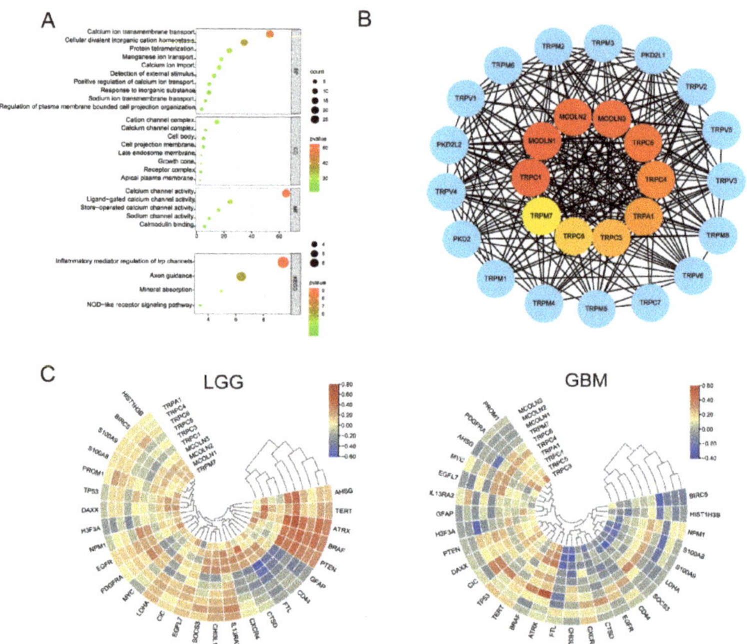

Figure 2. GO and KEGG enrichment analysis and hub genes of the TRP family in glioma. (**A**) The bubble plot shows the most enriched GO and KEGG pathways of target genes. Three factors were considered in the GO enrichment analysis, including BP, CC, and MF. Additionally, the most significant KEGG pathways involved inflammatory and their signaling pathways. (**B**) By using CytoHubba based on the MCC algorithm, the top 10 of the hub genes were identified in the center of the diagram. Additionally, the importance of genes was represented by different colors; red to yellow represent the decreasing importance of the hub genes, while blue was excluded from hub genes. (**C**) Heatmap representing the correlations between hub genes of TRP family and 28 biomarkers in glioma.

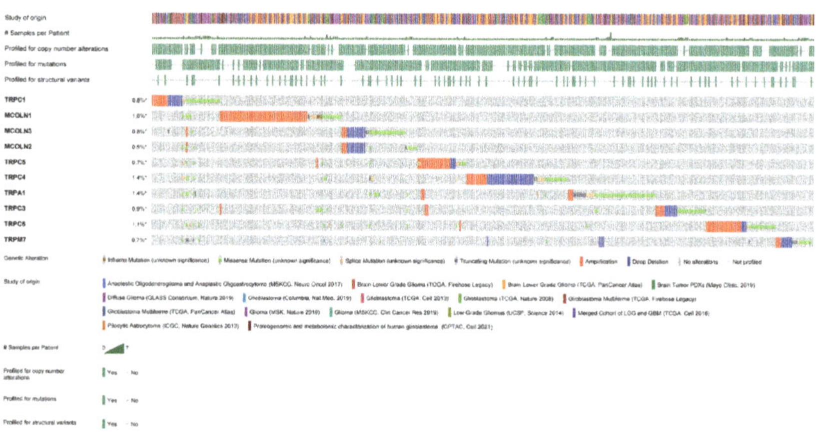

Figure 3. Amplification, deletion, and mutation of hub genes of TRPs family in glioma. Genetic variations in hub genes of the TRPs family were detected using the cBioportal database.

By inputting the hub genes as target genes into the HPA database, available representative immunohistochemistry images of TRPC1, TRPC6, TRPM7, MCOLN1, MCOLN2, and MCOLN3 were outputted. As shown in Figure 4B, compared with HGG tissues, the expression of TRPC1 was significantly higher in LGG tissues, while the expression of TRPC6, MCOLN1, MCOLN2, and MCOLN3 was relatively low, and these results are consistent with the above tendency.

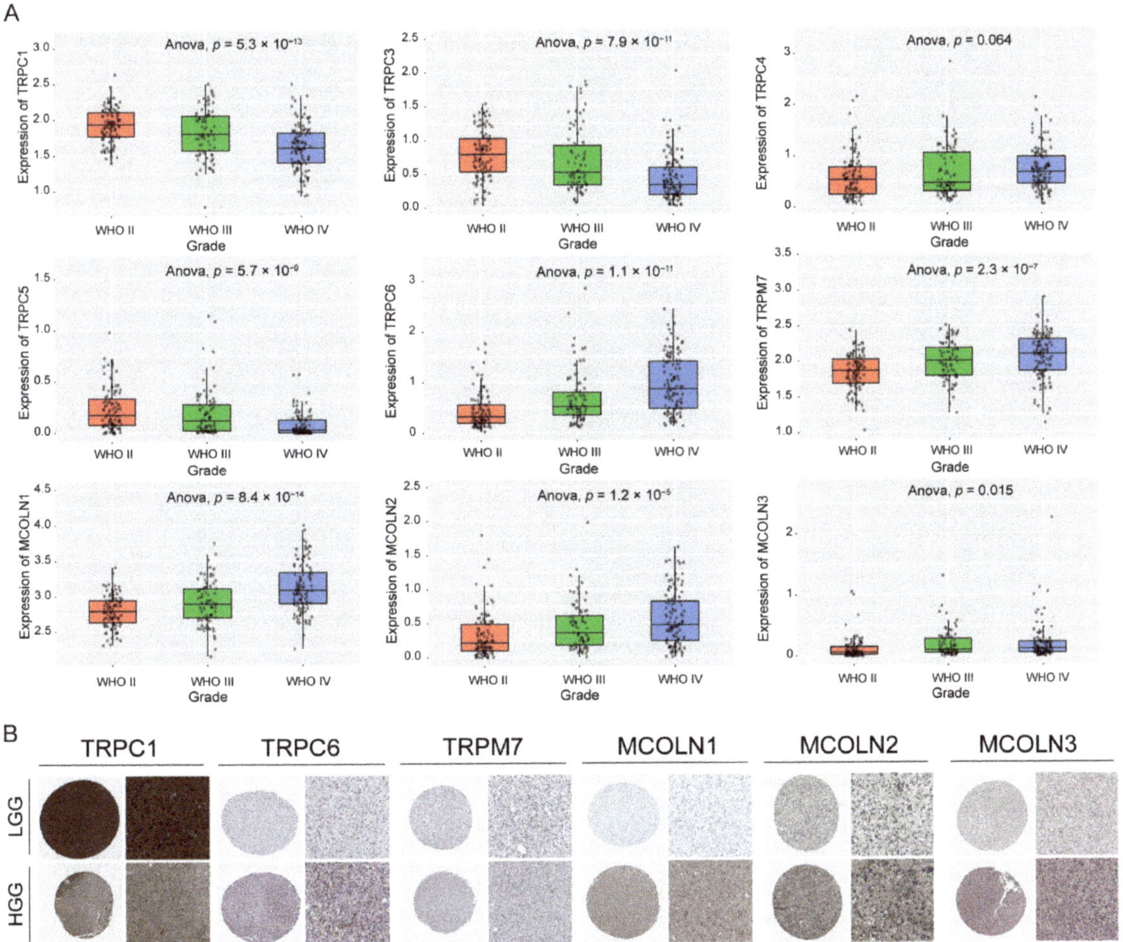

Figure 4. Expression of hub genes in the grade of glioma. (**A**). Validation of hub gene expression levels of TRPC1, MCOLN1, MCOLN2, MCOLN3, TRPC5, TRPC4, TRPC3, TRPC6, and TRPM7 in different grades of glioma using CGGA database. (**B**). Representative immunohistochemistry images of TRPC1, TRPC6, TRPM7, MCOLN1, MCOLN2, and MCOLN3 in HGG and LGG tissues derived from the HAP database. (HGG, high-grade glioma; LGG, low-grade glioma; HAP, Human Protein Atlas).

3.5. Survival Analysis of the Hub Genes in Glioma

To investigate the prognostic value of the above genes, we performed an analysis of OS and DFS. According to the results, we observed that analysis of the filtered hub genes through the above steps was further conducted by survival plots in the GEPIA database. As demonstrated in Figure 5, higher levels of TRPC1, TRPC3, and TRPC5 were significantly associated with longer OS time in glioma. Higher levels of TRPC4, TRPC6, MCOLN1, MCOLN2, and MCOLN3 were significantly associated with shorter OS times in glioma. Favorable DFS was significantly correlated with higher levels of TRPC1, TRPC3, and TRPC5 but significantly associated with lower levels of TRPC6, MCOLN1, MCOLN2, and MCOLN3. These findings were consistent with the obtained result of the expression profile of hub genes in the grade of glioma, which means that follow-up studies of TRPC1, MCOLN1, MCOLN2, MCOLN3, TRPC5, TRPC4, TRPC3, and TRPC6 will make more sense.

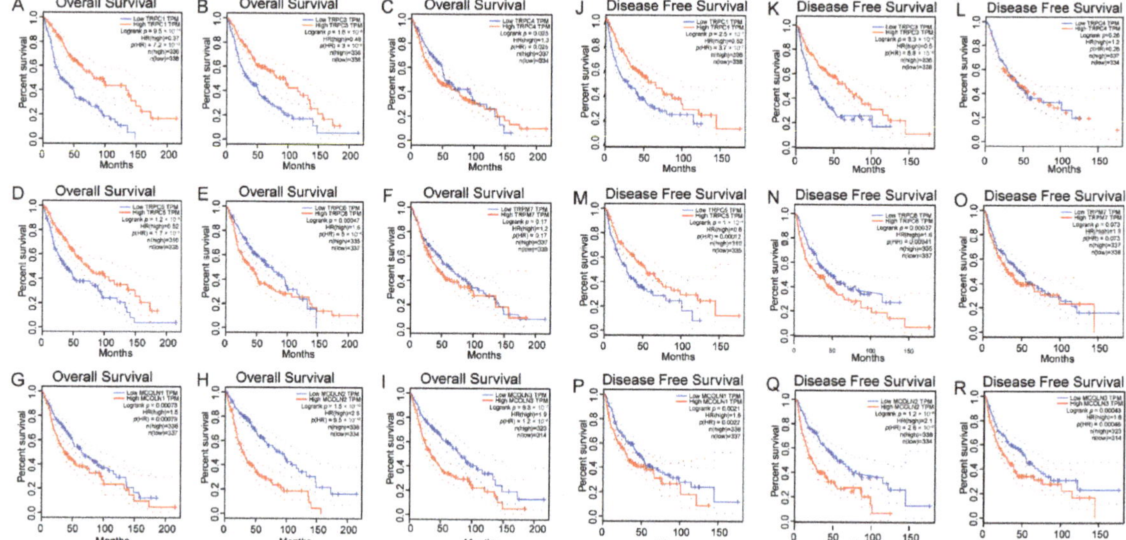

Figure 5. The dysregulation of hub genes of the TRPs family was correlated to OS and DFS time in glioma. (**A–I**) higher expression levels of TRPC1 (**A**), TRPC3 (**B**), and TRPC5 (**D**) were significantly associated with better OS time in glioma. Additionally, higher levels of TRPC4 (**C**), TRPC6 (**E**), MCOLN1 (**G**), MCOLN2 (**H**), and MCOLN3 (**I**) were significantly associated with worse OS time in glioma. (**J–R**) Similarly, higher levels of TRPC1 (**J**), TRPC3 (**K**), and TRPC5 (**M**) were significantly associated with better DFS time in glioma. Additionally, higher levels of TRPC6 (**N**), MCOLN1 (**P**), MCOLN2 (**Q**), and MCOLN3.

3.6. Expression Difference of Hub Genes in IDH Status, 1p/19q Codeletion Status of Glioma

By using the CGGA database, the expression levels of TRPC1, TRPC3, TRPC4, TRPC5, TRPC6, MCOLN1, MCOLN2, and MCOLN3 with IDH mutation status and 1p/19q codeletion status were compared. The expression levels of TRPC1 and TRPC3 were significantly higher in the 1p/19q codeletion status, while the gene expression levels of TRPC6, MCOLN1, MCOLN2, and MCOLN3 were significantly higher in the 1p/19q non-codeletion status (Figure 6B). IDH, a vital enzyme that participates in several essential metabolic processes, frequently exhibits a mutation status in human malignancies. A previous study indicated that patients with glioma with IDH mutation had a better OS time than those patients with wild-type IDH [42]. Similarly, the codeletion of chromosome status in glioma was associated with longer OS and DFS times, which means that 1p/19q codeletion status in patients with glioma could be used to predict a better prognosis [43]. The results demonstrated that the gene expression levels of TRPC1 and TRPC3 were significantly higher in IDH mutation status, while the gene expression levels of TRPC4, TRPC6, MCOLN1, MCOLN2, and MCOLN3 were significantly higher in IDH wild-type status (Figure 6A). The results were consistent with the survival analysis and expression differences above.

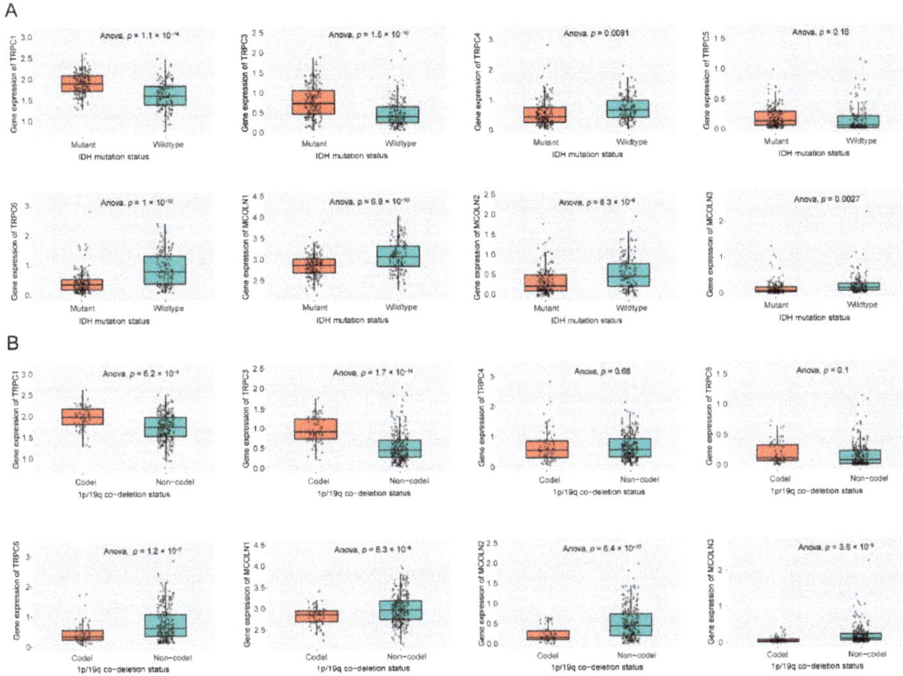

Figure 6. Expression of hub genes in IDH mutation status and 1p/19q codeletion status of glioma. (**A**). Validation of hub gene expression levels of TRPC1, MCOLN1, MCOLN2, MCOLN3, TRPC5, TRPC4, TRPC3, and TRPC6 in different IDH mutations status of glioma using CGGA database. (**B**). Validation of hub gene expression levels of TRPC1, MCOLN1, MCOLN2, MCOLN3, TRPC5, TRPC4, TRPC3, and TRPC6 in different 1p/19q codeletion status of glioma using CGGA database. The bottom and top of the boxes are the 25th and 75th percentiles, interquartile range. The statistical difference between the two subgroups was compared through the Kruskal–Wallis test, respectively.

3.7. Hub Genes Are Associated with Immune Activation in Glioma

By using the TIMER 2.0 database and TBtools software, we revealed the correlation of hub genes with the expression levels of immune cell infiltration, immune cell markers, and immune checkpoint molecules. The results that hub genes correlated with immune cell infiltration, immune cell markers, and immune checkpoint molecules in GBM and LGG are presented in Figure 7. As shown in Figure 7, a majority of hub genes were positively correlated with molecular markers.

3.8. Correlations between Hub Genes and the Expression of Immunoregulators in Glioma

By calculating the correlations between the expression of hub genes and immunoregulators, the effects of hub genes on the tumor immune response were explored (Figure 8). The results indicated that hub genes and the expression of immune regulators exhibited obvious correlations. TRP family genes have been shown to play key roles in inflammatory and immune cells by affecting cytokine production, cell differentiation, and cytotoxicity. In this section, we revealed that the expression of hub genes is correlated with immune activation and immunoregulators in glioma, which indicated that TRP family genes may serve as putative drug targets in glioma.

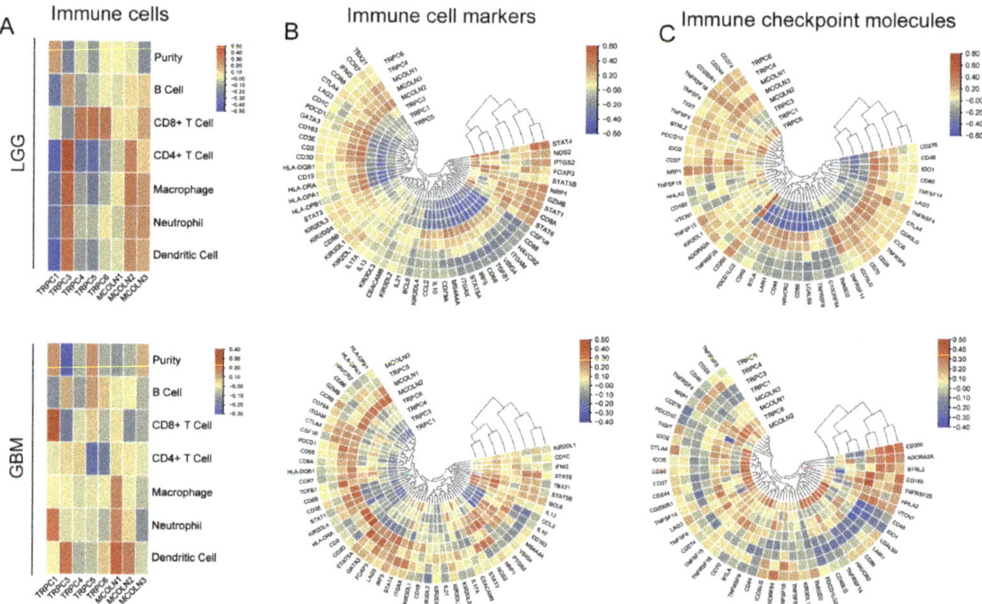

Figure 7. Hub genes are associated with immune activation in glioma. Heatmap represents the correlations between hub genes of TRPs family and 6 infiltrating immune cells (**A**), 56 immune cell markers (**B**), and 47 immune checkpoint molecules (**C**) in GBM and LGG.

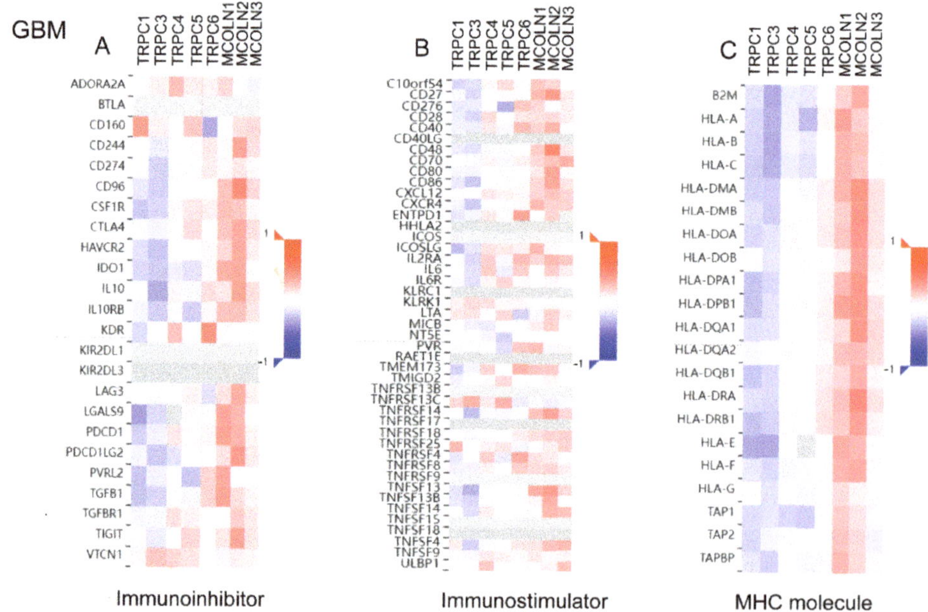

Figure 8. *Cont.*

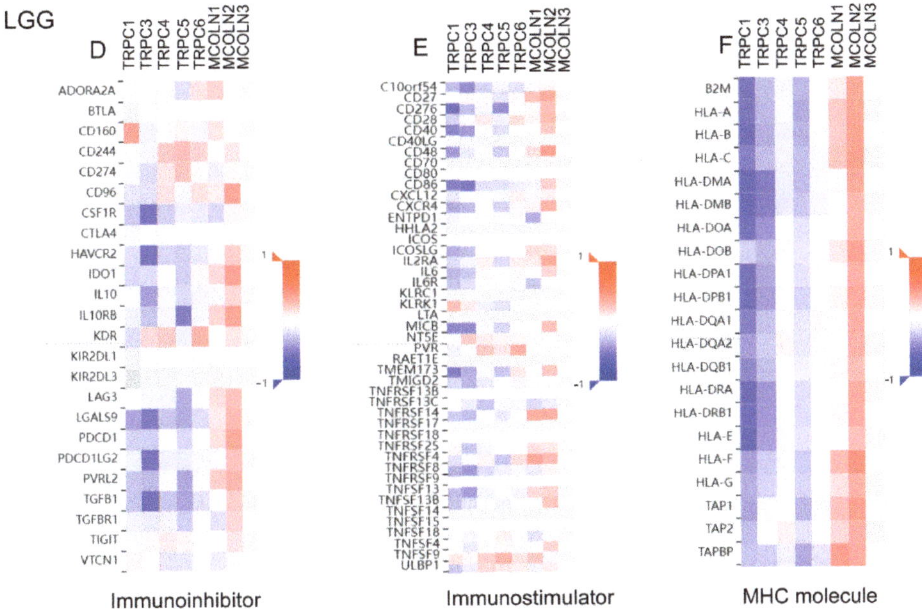

Figure 8. Correlations Between Hub Genes and the Expression of Immunoregulators in Glioma. The correlation between the expression of TRPC1, TRPC3, TRPC4, TRPC5, TRPC6, MCOLN1, MCOLN2, and MCOLN3 and immunoinhibitors (**A**,**D**), immunostimulators (**B**,**E**), and MHC molecules (**C**,**F**) was calculated based on the TISIDB database.

4. Discussion

Several studies have shed light on the role of TRP family genes in glioma in the past few years. In previous studies, the high expression of TRPC1 promoted tumor cell growth and migration through the mechanisms of cytokinesis and chemotaxis, respectively [44,45]. TRPC6 enhances cell proliferation, tumor growth, and angiogenesis, which is essential for glioma [46]. It has been shown that TRPML2 plays an important role in the progression of gliomas through the PI3K/AKT and ERK1/2 pathways [47]. In contrast, TRPM7 suppresses glioma progression through miR-28-5p regulation in GBM cells with subsequent effects on cell proliferation and invasion [48]. TRPML1, mediating autophagic cell death and exerting its cytotoxic effects in glioma, showed a protective effect against glioma progression [49,50]. Nevertheless, there remains a lack of wide analysis of the correlation between TRP family genes and glioma progression. Our study found significant prognostic value for TRP family genes in glioma by analyzing multiple datasets. Additionally, we determined whether TRP family genes affect antitumor immunity.

Our research systematically identified the most promising TRP family genes, which have the potential to be used as potential biomarkers in glioma prognosis and immunotherapy. We found that aberrations in TRP family genes in the genome were significantly correlated with prognosis in human glioma. Then, we performed functional enrichment analysis and found the top 10 hub genes of TRP family genes in glioma. The results showed that TRP family genes are mainly related to calcium ion transmembrane transport, cellular divalent inorganic cation homeostasis, and protein tetramerization in biological processes. In the cellular component category, TRP family genes were mainly enriched in cation channel complexes, calcium channel complexes, and the cell body. In molecular function, TRP family genes mainly correlate to calcium channel activity, ligand-gated calcium channel activity, and store-operated calcium channel activity. In KEGG-enriched terms, inflammatory mediator regulation of TRP channels displayed the highest correlation with TRP family genes. According to the results, we hypothesized that TRP family genes are highly associated with the immune response in glioma. Then, we found the top 10 hub genes in TRP family genes, which could be important genes correlated with immunotherapy and used as biomarkers. A previous study proved that the clinical value of TRPC1, TRPC6, TRPML1, and TRPML2 plays key roles in the diagnosis and prognosis of glioma patients. In our present study, we revealed that several novel TRP family genes may play an important role in glioma progression and be potential therapeutic targets.

Then, we found that aberrations in hub genes in the genome were correlated with prognosis in glioma. As shown in Figure 3, we revealed that all of the hub genes were amplified, deleted, and mutated in glioma, with MCOLN1 exhibiting the highest incidence rate (1.6%). The new studies indicate that MCOLN1/TRPML1 inhibits autophagy by inducing ROS-mediated TP53/p53 pathways, which inhibit cancer metastasis [51]. To verify the correlation between hub gene expression and malignancy in glioma, we investigated the expression of hub genes in glioma with different grades and explored their correlation with the time of OS and DFS. Our results indicated that all hub genes were differentially expressed in glioma grade except TRPA1. It is worth mentioning that although the expression of TRPV4 in different grades of glioma did not meet the inclusion criteria ($p > 0.05$), the difference in expression in OS time was significant. According to the screening criterion, the top 10 hub genes except TRPA1 and TRPM7 were finalized as reference genes for the follow-up study.

Among the eight final identified hub genes, higher expression levels of TRPC1, TRPC3, and TRPC5 were significantly associated with longer OS time in glioma, while higher expression levels of TRPC4, TRPC6, MCOLN1, MCOLN2, MCOLN3 were significantly associated with shorter OS time. In previous reports, TRPC1, TRPC3, TRPC6, TRPML1, and TRPML2 were proven to play crucial roles in glioma progression. Impairing TRPC6 activity in human glioma cells could reduce human xenograft growth in immunocompromised mice [52]. The expression level of TRPML2 was dramatically augmented in high-grade glioblastoma cell lines [47].

To evaluate whether hub genes are related to glioma patient prognosis, we further explored hub gene expression in patients with IDH mutations and 1p/19q codeletions. Previous studies indicated that patients with glioma with IDH mutation or 1p19q codeletion have a more favorable prognosis than patients with wild-type IDH or without 1p/19q codeletion [53,54]. In our present study, the results showed that the expression of TRPC1 and TRPC3 was significantly increased in glioma with IDH mutation status and 1p/19q codeletion status, while the expression of TRPC6, MCOLN1, MCOLN2, and MCOLN3 was significantly decreased. These results indicate that higher expression of TRPC1 and TRPC3 correlates with a better prognosis in patients with glioma, while higher expression of TRPC6, MCOLN1, MCOLN2, and MCOLN3 correlates with a worse prognosis in patients with glioma. The outcomes are consistent with previous results.

In recent years, much progress in immunotherapy has been made, and immunotherapy has promising prospects [55]. Previous studies have shown that immune cells, such as T cells, are closely correlated with tumor progression and clinical prognosis in patients and the response to immunotherapy in both humans and mice [56]. TRP family genes have been shown to play key roles in inflammatory and immune cells by affecting cytokine production, cell differentiation, and cytotoxicity [57]. However, few studies have clarified the relationship between TRP family genes and tumor immunotherapeutic targets. In our present study, we found that the expression of TRP family genes is strongly associated with a variety of cancers, especially glioma.

As shown in our previous study, we found that the TRP gene family was enriched in the inflammatory response. Previous studies have shown that the prognosis of glioma is related to immune cell infiltration, immune cell markers, and immune checkpoint molecules [2,22,58–60]. Some immunotherapy approaches, such as immune checkpoint blockade, have been successfully applied in organisms [61]. As a further demonstration of the predictive value of hub genes for immune therapy response in glioma, a heatmap was created to represent the correlation between hub genes and immune markers. According to the correlation between hub genes and immune biomarkers, we can predict the immunotherapy effects. In addition, we further validated the correlation between hub genes and the level of immune cell infiltration in cancer from TISIDB. The results were consistent with the conclusion beforehand.

5. Conclusions

In summary, the study identified the dysregulation of eight glioma-associated hub genes of the TRP family through bioinformatics analysis, which is correlated with the prognosis of glioma patients. Thus, it can be used as a prognostic biomarker. Further, we showed that the expression of hub genes is correlated with immune activation and immunoregulators in glioma. Our study indicated that TRP family genes may work as putative drug targets in glioma. We should be aware of several limitations of this study. First of all, the number of validation databases in this article is insufficient. A future study will need to examine other public databases for verification to confirm the results obtained in this study. Secondly, there is a lack of clinical practice in our conclusion, so further clinical trials are needed to confirm whether TRP family genes can be used as new glioma treatment targets. Thirdly, the current study does not yet prove that the hub genes can be used as an independent prognostic factor, which means that further trials are needed. In conclusion, our study identified characteristics of TRP family genes, which can be used to predict prognosis for glioma patients and are associated with immunological response, providing a new potential treatment strategy for glioma patients.

Author Contributions: C.F., H.X., Y.L. and C.H. were involved in data analysis and interpretation. Y.X., L.Y., X.W. and Z.Z. typeset pictures. A.Z., A.S. and M.L. designed the experiment, interpreted the data, and wrote the manuscript. All authors have read and agreed to the published version of the manuscript.

Funding: This work was supported by the Natural science foundation of Shanghai (18ZR1430400) and the Zhejiang Provincial Natural Science Foundation of China (LY22H090020).

Institutional Review Board Statement: Not applicable.

Informed Consent Statement: Not applicable.

Data Availability Statement: The original contributions presented in the study are included in the article, further inquiries can be directed to the corresponding authors.

Conflicts of Interest: The authors declare no conflict of interest.

References

1. Ostrom, Q.T.; Cioffi, G.; Waite, K.; Kruchko, C.; Barnholtz-Sloan, J.S. CBTRUS Statistical Report: Primary Brain and Other Central Nervous System Tumors Diagnosed in the United States in 2014–2018. *Neuro Oncol.* **2021**, *23*, iii1–iii105. [CrossRef] [PubMed]
2. Boyd, N.H.; Tran, A.N.; Bernstock, J.D.; Etminan, T.; Jones, A.B.; Gillespie, G.Y.; Friedman, G.K.; Hjelmeland, A.B. Glioma stem cells and their roles within the hypoxic tumor microenvironment. *Theranostics* **2021**, *11*, 665–683. [CrossRef] [PubMed]
3. Wesseling, P.; Capper, D. WHO 2016 Classification of gliomas. *Neuropathol. Appl. Neurobiol.* **2018**, *44*, 139–150. [CrossRef] [PubMed]
4. Molloy, A.R.; Najac, C.; Viswanath, P.; Lakhani, A.; Subramani, E.; Batsios, G.; Radoul, M.; Gillespie, A.M.; Pieper, R.O.; Ronen, S.M. MR-detectable metabolic biomarkers of response to mutant IDH inhibition in low-grade glioma. *Theranostics* **2020**, *10*, 8757–8770. [CrossRef]
5. Baker, S.D.; Wirth, M.; Statkevich, P.; Reidenberg, P.; Alton, K.; Sartorius, S.E.; Dugan, M.; Cutler, D.; Batra, V.; Grochow, L.B.; et al. Absorption, metabolism, and excretion of 14C-temozolomide following oral administration to patients with advanced cancer. *Clin. Cancer Res.* **1999**, *5*, 309–317.
6. Laquintana, V.; Trapani, A.; Denora, N.; Wang, F.; Gallo, J.M.; Trapani, G. New strategies to deliver anticancer drugs to brain tumors. *Expert Opin. Drug Deliv.* **2009**, *6*, 1017–1032. [CrossRef]
7. Bush, N.A.O.; Chang, S.M.; Berger, M.S. Current and future strategies for treatment of glioma. *Neurosurg. Rev.* **2016**, *40*, 1–14. [CrossRef]
8. Zhou, Q.; Yan, X.; Zhu, H.; Xin, Z.; Zhao, J.; Shen, W.; Yin, W.; Guo, Y.; Xu, H.; Zhao, M.; et al. Identification of three tumor antigens and immune subtypes for mRNA vaccine development in diffuse glioma. *Theranostics* **2021**, *11*, 9775–9790. [CrossRef]

9. Korman, A.J.; Peggs, K.S.; Allison, J.P. Checkpoint Blockade in Cancer Immunotherapy. *Adv. Immunol.* **2006**, *90*, 297–339. [CrossRef]

10. Wainwright, D.; Chang, A.L.; Dey, M.; Balyasnikova, I.V.; Kim, C.K.; Tobias, A.; Cheng, Y.; Kim, J.W.; Qiao, J.; Zhang, L.; et al. Durable Therapeutic Efficacy Utilizing Combinatorial Blockade against IDO, CTLA-4, and PD-L1 in Mice with Brain Tumors. *Clin. Cancer Res.* **2014**, *20*, 5290–5301. [CrossRef]

11. Santoni, G.; Maggi, F.; Morelli, M.B.; Santoni, M.; Marinelli, O. Transient Receptor Potential Cation Channels in Cancer Therapy. *Med. Sci.* **2019**, *7*, 108. [CrossRef] [PubMed]

12. Prevarskaya, N.; Zhang, L.; Barritt, G. TRP channels in cancer. *Biochim. Biophys. Acta* **2007**, *1772*, 937–946. [CrossRef]

13. Stokłosa, P.; Borgström, A.; Kappel, S.; Peinelt, C. TRP Channels in Digestive Tract Cancers. *Int. J. Mol. Sci.* **2020**, *21*, 1877. [CrossRef] [PubMed]

14. Gkika, D.; Prevarskaya, N. Molecular mechanisms of TRP regulation in tumor growth and metastasis. *Biochim. Biophys. Acta* **2009**, *1793*, 953–958. [CrossRef]

15. Gil-Kulik, P.; Dudzińska, E.; Radzikowska-Büchner, E.; Wawer, J.; Jojczuk, M.; Nogalski, A.; Wawer, G.A.; Feldo, M.; Kocki, W.; Cioch, M.; et al. Different regulation of PARP1, PARP2, PARP3 and TRPM2 genes expression in acute myeloid leukemia cells. *BMC Cancer* **2020**, *20*, 435. [CrossRef]

16. Wang, Y.; Qi, Y.X.; Qi, Z.; Tsang, S.Y. TRPC3 Regulates the Proliferation and Apoptosis Resistance of Triple Negative Breast Cancer Cells through the TRPC3/RASA4/MAPK Pathway. *Cancers* **2019**, *11*, 558. [CrossRef] [PubMed]

17. Chinigò, G.; Castel, H.; Chever, O.; Gkika, D. TRP Channels in Brain Tumors. *Front. Cell Dev. Biol.* **2021**, *9*, 617801. [CrossRef]

18. Nabissi, M.; Morelli, M.B.; Amantini, C.; Farfariello, V.; Ricci-Vitiani, L.; Caprodossi, S.; Arcella, A.; Santoni, M.; Giangaspero, F.; De Maria, R.; et al. TRPV2 channel negatively controls glioma cell proliferation and resistance to Fas-induced apoptosis in ERK-dependent manner. *Carcinogenesis* **2010**, *31*, 794–803. [CrossRef]

19. Bomben, V.C.; Sontheimer, H. Disruption of transient receptor potential canonical channel 1 causes incomplete cytokinesis and slows the growth of human malignant gliomas. *Glia* **2010**, *58*, 1145–1156. [CrossRef]

20. Chigurupati, S.; Venkataraman, R.; Barrera, D.; Naganathan, A.; Madan, M.; Paul, L.; Pattisapu, J.V.; Kyriazis, G.; Sugaya, K.; Bushnev, S.; et al. Receptor Channel TRPC6 Is a Key Mediator of Notch-Driven Glioblastoma Growth and Invasiveness. *Cancer Res.* **2009**, *70*, 418–427. [CrossRef]

21. Ambrosino, P.; Soldovieri, M.V.; Russo, C.; Taglialatela, M. Activation and desensitization of TRPV1 channels in sensory neurons by the PPARα agonist palmitoylethanolamide. *Br. J. Pharmacol.* **2012**, *168*, 1430–1444. [CrossRef] [PubMed]

22. Lee, S.H.; Tonello, R.; Im, S.-T.; Jeon, H.; Park, J.; Ford, Z.; Davidson, S.; Kim, Y.H.; Park, C.-K.; Berta, T. Resolvin D3 controls mouse and human TRPV1-positive neurons and preclinical progression of psoriasis. *Theranostics* **2020**, *10*, 12111–12126. [CrossRef] [PubMed]

23. Bang, S.; Yoo, S.; Yang, T.; Cho, H.; Hwang, S. Nociceptive and pro-inflammatory effects of dimethylallyl pyrophosphate via TRPV4 activation. *Br. J. Pharmacol.* **2012**, *166*, 1433–1443. [CrossRef] [PubMed]

24. Zhao, P.; Lieu, T.; Barlow, N.; Metcalf, M.; Veldhuis, N.; Jensen, D.D.; Kocan, M.; Sostegni, S.; Haerteis, S.; Baraznenok, V.; et al. Cathepsin S Causes Inflammatory Pain via Biased Agonism of PAR2 and TRPV4. *J. Biol. Chem.* **2014**, *289*, 27215–27234. [CrossRef] [PubMed]

25. Multhoff, G.; Molls, M.; Radons, J. Chronic Inflammation in Cancer Development. *Front. Immunol.* **2012**, *2*, 98. [CrossRef] [PubMed]

26. Ravindranathan, D.; Master, V.; Bilen, M. Inflammatory Markers in Cancer Immunotherapy. *Biology* **2021**, *10*, 325. [CrossRef]

27. Unberath, P.; Knell, C.; Prokosch, H.-U.; Christoph, J. Developing New Analysis Functions for a Translational Research Platform: Extending the cBioPortal for Cancer Genomics. *Stud. Health Technol. Inform.* **2019**, *258*, 46–50. [CrossRef]

28. Gene Ontology Consortium. Gene Ontology Consortium: Going forward. *Nucleic Acids Res.* **2015**, *43*, D1049–D1056. [CrossRef]

29. Kanehisa, M.; Furumichi, M.; Tanabe, M.; Sato, Y.; Morishima, K. KEGG: New perspectives on genomes, pathways, diseases and drugs. *Nucleic Acids Res.* **2017**, *45*, D353–D361. [CrossRef]

30. Zhao, Z.; Zhang, K.-N.; Wang, Q.; Li, G.; Zeng, F.; Zhang, Y.; Wu, F.; Chai, R.; Wang, Z.; Zhang, C.; et al. Chinese Glioma Genome Atlas (CGGA): A Comprehensive Resource with Functional Genomic Data from Chinese Glioma Patients. *Genom. Proteom. Bioinform.* **2021**, *19*, 1–12. [CrossRef]

31. Asplund, A.; Edqvist, P.-H.D.; Schwenk, J.M.; Pontén, F. Antibodies for profiling the human proteome-The Human Protein Atlas as a resource for cancer research. *Proteomics* **2012**, *12*, 2067–2077. [CrossRef] [PubMed]

32. Gu, Y.; Li, X.; Bi, Y.; Zheng, Y.; Wang, J.; Li, X.; Huang, Z.; Chen, L.; Huang, Y.; Huang, Y. CCL14 is a prognostic biomarker and correlates with immune infiltrates in hepatocellular carcinoma. *Aging* **2020**, *12*, 784–807. [CrossRef] [PubMed]

33. Peng, L.; Chen, Z.; Chen, Y.; Wang, X.; Tang, N. *MIR155HG* is a prognostic biomarker and associated with immune infiltration and immune checkpoint molecules expression in multiple cancers. *Cancer Med.* **2019**, *8*, 7161–7173. [CrossRef] [PubMed]

34. Tu, L.; Guan, R.; Yang, H.; Zhou, Y.; Hong, W.; Ma, L.; Zhao, G.; Yu, M. Assessment of the expression of the immune checkpoint molecules PD-1, CTLA4, TIM-3 and LAG-3 across different cancers in relation to treatment response, tumor-infiltrating immune cells and survival. *Int. J. Cancer* **2020**, *147*, 423–439. [CrossRef] [PubMed]

35. Li, T.; Fan, J.; Wang, B.; Traugh, N.; Chen, Q.; Liu, J.S.; Li, B.; Liu, X.S. TIMER: A Web Server for Comprehensive Analysis of Tumor-Infiltrating Immune Cells. *Cancer Res.* **2017**, *77*, e108–e110. [CrossRef]

36. Ru, B.; Wong, C.N.; Tong, Y.; Zhong, J.Y.; Zhong, S.S.W.; Wu, W.C.; Chu, K.C.; Wong, C.Y.; Lau, C.Y.; Chen, I.; et al. TISIDB: An integrated repository portal for tumor–immune system interactions. *Bioinformatics* **2019**, *35*, 4200–4202. [CrossRef]

37. Dong, H.; Shim, K.-N.; Li, J.M.J.; Estrema, C.; Ornelas, T.A.; Nguyen, F.; Liu, S.; Ramamoorthy, S.L.; Ho, S.; Carethers, J.M.; et al. Molecular mechanisms underlying Ca^{2+}-mediated motility of human pancreatic duct cells. *Am. J. Physiol. Cell Physiol.* **2010**, *299*, C1493–C1503. [CrossRef]

38. Gao, S.-L.; Kong, C.-Z.; Zhang, Z.; Li, Z.-L.; Bi, J.-B.; Liu, X.-K. TRPM7 is overexpressed in bladder cancer and promotes proliferation, migration, invasion and tumor growth. *Oncol. Rep.* **2017**, *38*, 1967–1976. [CrossRef]

39. Nakashima, S.; Shiozaki, A.; Ichikawa, D.; Hikami, S.; Kosuga, T.; Konishi, H.; Komatsu, S.; Fujiwara, H.; Okamoto, K.; Kishimoto, M.; et al. Transient Receptor Potential Melastatin 7 as an Independent Prognostic Factor in Human Esophageal Squamous Cell Carcinoma. *Anticancer Res.* **2017**, *37*, 1161–1168. [CrossRef]

40. Li, X.; Zhang, Q.; Fan, K.; Li, B.; Li, H.; Qi, H.; Guo, J.; Cao, Y.; Sun, H. Overexpression of TRPV3 Correlates with Tumor Progression in Non-Small Cell Lung Cancer. *Int. J. Mol. Sci.* **2016**, *17*, 437. [CrossRef]

41. Çiğ, B.; Derouiche, S.; Jiang, L.-H. Editorial: Emerging Roles of TRP Channels in Brain Pathology. *Front. Cell Dev. Biol.* **2021**, *9*, 705196. [CrossRef] [PubMed]

42. Yan, H.; Parsons, D.W.; Jin, G.; McLendon, R.; Rasheed, B.A.; Yuan, W.; Kos, I.; Batinic-Haberle, I.; Jones, S.; Riggins, G.J.; et al. IDH1 and IDH2 mutations in gliomas. *N. Engl. J. Med.* **2009**, *360*, 765–773. [CrossRef]

43. Jenkins, R.B.; Blair, H.; Ballman, K.V.; Giannini, C.; Arusell, R.M.; Law, M.; Flynn, H.; Passe, S.; Felten, S.; Brown, P.D.; et al. A t(1;19)(q10;p10) Mediates the Combined Deletions of 1p and 19q and Predicts a Better Prognosis of Patients with Oligodendroglioma. *Cancer Res.* **2006**, *66*, 9852–9861. [CrossRef] [PubMed]

44. Bomben, V.C.; Turner, K.L.; Barclay, T.-T.C.; Sontheimer, H. Transient receptor potential canonical channels are essential for chemotactic migration of human malignant gliomas. *J. Cell. Physiol.* **2011**, *226*, 1879–1888. [CrossRef]

45. Bomben, V.C.; Sontheimer, H.W. Inhibition of transient receptor potential canonical channels impairs cytokinesis in human malignant gliomas. *Cell Prolif.* **2008**, *41*, 98–121. [CrossRef] [PubMed]

46. Neimark, M.A.; Konstas, A.-A.; Laine, A.F.; Pile-Spellman, J. Integration of jugular venous return and circle of Willis in a theoretical human model of selective brain cooling. *J. Appl. Physiol.* **2007**, *103*, 1837–1847. [CrossRef] [PubMed]

47. Morelli, M.B.; Nabissi, M.; Amantini, C.; Tomassoni, D.; Rossi, F.; Cardinali, C.; Santoni, M.; Arcella, A.; Oliva, M.A.; Santoni, A.; et al. Overexpression of transient receptor potential mucolipin-2 ion channels in gliomas: Role in tumor growth and progression. *Oncotarget* **2016**, *7*, 43654–43668. [CrossRef]

48. Wan, J.; Guo, A.A.; Chowdhury, I.; Guo, S.; Hibbert, J.; Wang, G.; Liu, M. TRPM7 Induces Mechanistic Target of Rap1b Through the Downregulation of miR-28-5p in Glioma Proliferation and Invasion. *Front. Oncol.* **2019**, *9*, 1413. [CrossRef]

49. Morelli, M.B.; Amantini, C.; Tomassoni, D.; Nabissi, M.; Arcella, A.; Santoni, G. Transient Receptor Potential Mucolipin-1 Channels in Glioblastoma: Role in Patient's Survival. *Cancers* **2019**, *11*, 525. [CrossRef]

50. Zhang, X.; Cheng, X.; Yu, L.; Yang, J.; Calvo, R.; Patnaik, S.; Hu, X.; Gao, Q.; Yang, M.; Lawas, M.; et al. MCOLN1 is a ROS sensor in lysosomes that regulates autophagy. *Nat. Commun.* **2016**, *7*, 12109. [CrossRef]

51. Xing, Y.; Wei, X.; Liu, Y.; Wang, M.M.; Sui, Z.; Wang, X.; Zhu, W.; Wu, M.; Lu, C.; Fei, Y.H.; et al. Autophagy inhibition mediated by MCOLN1/TRPML1 suppresses cancer metastasis via regulating a ROS-driven TP53/p53 pathway. *Autophagy* **2021**, 1–23. [CrossRef] [PubMed]

52. Ding, X.; He, Z.; Zhou, K.; Cheng, J.; Yao, H.; Lu, D.; Cai, R.; Jin, Y.; Dong, B.; Xu, Y.; et al. Essential role of TRPC6 channels in G2/M phase transition and development of human glioma. *J. Natl. Cancer Inst.* **2010**, *102*, 1052–1068. [CrossRef] [PubMed]

53. McNamara, M.G.; Jiang, H.; Lim-Fat, M.J.; Sahebjam, S.; Kiehl, T.-R.; Karamchandani, J.; Coire, C.; Chung, C.; Millar, B.-A.; Laperriere, N.; et al. Treatment Outcomes in 1p19q Co-deleted/Partially Deleted Gliomas. *Can. J. Neurol. Sci.* **2017**, *44*, 288–294. [CrossRef] [PubMed]

54. Dang, L.; Jin, S.; Su, S.M. IDH mutations in glioma and acute myeloid leukemia. *Trends Mol. Med.* **2010**, *16*, 387–397. [CrossRef]

55. Ascierto, P.A.; Bifulco, C.; Buonaguro, L.; Emens, L.A.; Ferris, R.L.; Fox, B.A.; Delgoffe, G.M.; Galon, J.; Gridelli, C.; Merlano, M.; et al. Perspectives in immunotherapy: Meeting report from the "Immunotherapy Bridge 2018" (28–29 November, 2018, Naples, Italy). *J. Immunother. Cancer* **2019**, *7*, 332. [CrossRef]

56. Lanitis, E.; Dangaj, D.; Irving, M.; Coukos, G. Mechanisms regulating T-cell infiltration and activity in solid tumors. *Ann. Oncol.* **2017**, *28*, xii18–xii32. [CrossRef]

57. Parenti, A.; De Logu, F.; Geppetti, P.; Benemei, S. What is the evidence for the role of TRP channels in inflammatory and immune cells? *Br. J. Pharmacol.* **2016**, *173*, 953–969. [CrossRef]

58. Zhou, Q.; Yan, X.; Liu, W.; Yin, W.; Xu, H.; Cheng, D.; Jiang, X.; Ren, C. Three Immune-Associated Subtypes of Diffuse Glioma Differ in Immune Infiltration, Immune Checkpoint Molecules, and Prognosis. *Front. Oncol.* **2020**, *10*, 586019. [CrossRef]

59. Huang, S.; Song, Z.; Zhang, T.; He, X.; Huang, K.; Zhang, Q.; Shen, J.; Pan, J. Identification of Immune Cell Infiltration and Immune-Related Genes in the Tumor Microenvironment of Glioblastomas. *Front. Immunol.* **2020**, *11*, 585034. [CrossRef]

60. Yang, M.; Tang, X.; Zhang, Z.; Gu, L.; Wei, H.; Zhao, S.; Zhong, K.; Mu, M.; Huang, C.; Jiang, C.; et al. Tandem CAR-T cells targeting CD70 and B7-H3 exhibit potent preclinical activity against multiple solid tumors. *Theranostics* **2020**, *10*, 7622–7634. [CrossRef]

61. Saha, D.; Martuza, R.L.; Rabkin, S.D. Macrophage Polarization Contributes to Glioblastoma Eradication by Combination Immunovirotherapy and Immune Checkpoint Blockade. *Cancer Cell* **2017**, *32*, 253–267.e5. [CrossRef] [PubMed]

Article

Immune Landscape in PTEN-Related Glioma Microenvironment: A Bioinformatic Analysis

Alice Giotta Lucifero [1] and Sabino Luzzi [1,2,*]

1   Neurosurgery Unit, Department of Clinical-Surgical, Diagnostic and Pediatric Sciences, University of Pavia, 27100 Pavia, Italy; alicelucifero@gmail.com
2   Neurosurgery Unit, Department of Surgical Sciences, Fondazione IRCCS Policlinico San Matteo, 27100 Pavia, Italy
*   Correspondence: sabino.luzzi@unipv.it

Abstract: Introduction: PTEN gene mutations are frequently found in the genetic landscape of high-grade gliomas since they influence cell proliferation, proangiogenetic pathways, and antitumoral immune response. The present bioinformatics analysis explores the PTEN gene expression profile in HGGs as a prognostic factor for survival, especially focusing on the related immune microenvironment. The effects of PTEN mutation on the susceptibility to conventional chemotherapy were also investigated. Methods: Clinical and genetic data of GBMs and normal tissue samples were acquired from The Cancer Genome Atlas (TCGA)-GBM and Genotype-Tissue Expression (GTEx) online databases, respectively. The genetic differential expressions were analyzed in both groups via the one-way ANOVA test. Kaplan–Meier survival curves were applied to estimate the overall survival (OS) and disease-free survival (DFS). The Genomics of Drug Sensitivity in Cancer platform was chosen to assess the response of PTEN-mutated GBMs to temozolomide (TMZ). $p < 0.05$ was fixed as statistically significant. On Tumor Immune Estimation Resource and Gene Expression Profiling Interactive Analysis databases, the linkage between immune cell recruitment and PTEN status was assessed through Spearman's correlation analysis. Results: PTEN was found mutated in 22.2% of the 617 TCGA-GBMs patients, with a higher log2-transcriptome per million reads compared to the GTEx group (255 samples). Survival curves revealed a worse OS and DFS, albeit not significant, for the high-PTEN profile GBMs. Spearman's analysis of immune cells demonstrated a strong positive correlation between the PTEN status and infiltration of T_{reg} ($\rho = 0.179$) and M2 macrophages ($\rho = 0.303$). The half-maximal inhibitor concentration of TMZ was proven to be lower for PTEN-mutated GBMs compared with PTEN wild-types. Conclusions: PTEN gene mutations prevail in GBMs and are strongly related to poor prognosis and least survival. The infiltrating immune lymphocytes T_{reg} and M2 macrophages populate the glioma microenvironment and control the mechanisms of tumor progression, immune escape, and sensitivity to standard chemotherapy. Broader studies are required to confirm these findings and turn them into new therapeutic perspectives.

Keywords: glioblastoma; high-grade gliomas; phosphatase and tensin homolog; PTEN; temozolomide; The Cancer Genome Atlas

Citation: Giotta Lucifero, A.; Luzzi, S. Immune Landscape in PTEN-Related Glioma Microenvironment: A Bioinformatic Analysis. *Brain Sci.* **2022**, *12*, 501. https://doi.org/10.3390/brainsci12040501

Academic Editor: Hailiang Tang

Received: 8 March 2022
Accepted: 12 April 2022
Published: 14 April 2022

Publisher's Note: MDPI stays neutral with regard to jurisdictional claims in published maps and institutional affiliations.

1. Introduction

High-grade gliomas (HGGs) are common neoplasms of the central nervous system accounting for 70% of brain tumors [1–5]. Glioblastoma (GBM) is the most lethal and represents 60% of newly diagnosed gliomas in the adult population [6,7]. The current standard of care in the management of HGGs is maximum surgical resection, adjuvant chemoradiation, and six cycles of temozolomide [8].

Despite advances in surgical techniques, diagnostics, and target therapeutic strategies, the 5-year survival rate persists under 10% and the median overall survival (OS) still ranges between 14 and 16 months [9]. The poor prognosis and high mortality rate of GBM are attributable to aberrant angiogenesis, extreme mitotic activity, immune escape mechanisms, and intrinsic genome-wide heterogeneity [10–17]. In 2021, Louis and colleagues published the fifth edition of the WHO classification of brain cancers, which reflects the advances in translational medicine, taxonomy, and genetics in neuro-oncology.

They reported a novel tumors nomenclature aimed at integrating histological features, key diagnostic genes, and molecular characteristics underlying oncogenesis [18].

Phosphatase and tensin homolog (PTEN), a tumor suppressor gene, is closely involved in cell translation, proliferation, and tumorigenesis [19–22]. PTEN protein blocks the intracellular pathways of phosphatidylinositol 3-kinase/AKT/mammalian target of rapamycin (PI3K/AKT/mTOR) via dephosphorylation of phosphatidylinositol-3-triphosphate, resulting in inhibition of the cell cycle [23–26].

Moreover, the PTEN pathway has a close interaction with the Wnt/β-catenin signals involved in embryogenesis and the determination of neural patterning [22].

Lack/mutation of the PTEN gene, found in 40% of GBMs, influences neurogenesis, and gliogenesis and heightens the DNA damage repairing and the malignant progression of brain tumors [20,25,27].

The prognostic significance of the PTEN gene is also related to the maintenance of the immune microenvironment [28]. Recent transcriptomic pieces of evidence support the correlation between PTEN mutation and the amendment of immune infiltrating cells expression [29,30]. The immune suppression mechanisms intrude on the host antitumor responses and are liable for the failure of conventional therapeutic approaches [31–35]. The number of reported studies on the immunogenomics and immunosuppressive microenvironment of HGGS opened the way for tailored treatments against glioma resilience [36–39]. Based on these assumptions, we conducted a bioinformatics analysis to examine the PTEN gene expression profile in HGGs as a prognostic factor for survival. We investigated the cluster of the immune infiltrates within the PTEN-related microenvironment intending to identify the contribution of each subpopulation to the immune escape mechanisms. The effects of PTEN mutation on the sensitivity to standard chemotherapy were also explored.

2. Materials and Methods

2.1. Data Acquisition

Transcriptomes, genetic, and clinical features of HGGs-patients were extracted from The Cancer Genome Atlas (TCGA)-GBM project (https://portal.gdc.cancer.gov) (accessed on 31 October 2021) [40]. The genetic data of normal brain tissue samples were acquired from the Genotype-Tissue Expression (GTEx) online database (https://gtexportal.org) (accessed on 1 November 2021).

2.2. Statistical Bioinformatics Analysis

R (https://www.r-project.org) (accessed on 7 March 2022) and Prism 5 (GraphPad Software, Inc., La Jolla, CA, USA) software were used for statistical analysis. Continuous and categorical variables were reported as mean and percentages, respectively.

Top mutation trends and gene patterns were estimated in the TCGA-GBM cohort, focusing on nucleotide variations of the PTEN gene. The differential expressions of PTEN mRNA levels were determined in TCGA-GBM and GTEx groups with the aim to assess the significance of the PTEN gene in the glioma genome compared to healthy brain tissue. A one-way ANOVA test was used for the analysis.

Kaplan–Meier survival curves were used to estimate the prognostic value of the PTEN gene mutations. The overall survival (OS), disease-free survival (DFS), and comparisons between the high- or low-PTEN mutation profile in TCGA-GBM patients were assessed using the log-rank test. Hazard ratios (HDs) were calculated with the Cox proportional risk regression model. The Genomics of Drug Sensitivity in Cancer (GDSC) database (https://www.cancerrxgene.org/) (accessed on 15 November 2021) was the source used for the appraisal of the chemotherapeutic response of PTEN-mutated HGGs [41]. The R software package "Prophetic" was used and the half-maximal inhibitor concentration

(IC50) of temozolomide (TMZ) was assessed by one-way ANOVA analysis [41]. *p*-value was set at <0.05 for all the tests.

2.3. Estimation of Immune Infiltrating Cells

On behalf of the glioma immune microenvironment, the T cells CD4+, CD8+, T_{reg}, NK cells, monocytes/macrophages, and tumor-infiltrating endothelial cells were considered in the TCGA-GBM cohort. Tumor Immune Estimation Resource 2.0 (TIMER2.0) was used to identify the correlation between mRNA PTEN expression and the transcriptional profile of each tumor immune cell in the TCGA-GBM project.

Assuming the purity adjustment, Spearman's correlation method was employed, where rho (ρ) > 0 and ρ < 0 denoted a positive and negative correlation between the variables, respectively. Based on the Gene Expression Profiling Interactive Analysis (GEPIA), the differential subexpression of mRNA PTEN in the immune subtypes, for both TCGA-GBM and GTEx samples, was clustered with the one-way ANOVA method, and $p < 0.05$ was assumed as statistical. The results were reported as boxplots.

3. Results

3.1. Demographics and Gene Mutation Profiles

The genetic and clinical data of 617 HGGs were collected by the TCGA-GBM project. The average patients' age was 58.8 ± 14 years; males were 59.5%, and 96.3% were Caucasian. All the tumors were supratentorial glioblastoma. Chemotherapy was administered as adjuvant treatment in 52.2% of patients, while 45% were treated with concomitant radiotherapy. The average follow-up was 14.7 months, and 69.2% of patients were dead. Overall data about TCGA-GBM patients are summarized in Table 1.

Table 1. Demographic, clinical, and histological data of TCGA-GBM patients.

Variable	Data
Case no.	617
Average pts. age (yrs ± SD)	58.8 ± 14
Sex	
Male no. (%)	367 (59.5)
Female no. (%)	230 (37.3)
NOS no. (%)	20 (3.2)
Ethnicity	
Caucasian no. (%)	594 (96.3)
African no. (%)	20 (3.2)
Asian no. (%)	3 (0.5)
Histological type	
Glioblastoma no. (%)	617 (100)
IDH wild-type ratio/IDH-mutant ratio	577:40
Primary site	
Brain no. (%)	617 (100)
Adjuvant treatment	
Pharmacotherapy no. (%)	322 (52.2)
Radiation Therapy no. (%)	278 (45)
NOS no. (%)	17 (2.8)

Table 1. *Cont.*

Variable	Data
Histological type	
Glioblastoma no. (%)	617 (100)
IDH wild-type ratio/IDH-mutant ratio	577:40
Primary site	
Brain no. (%)	617 (100)
Adjuvant treatment	
Pharmacotherapy no. (%)	322 (52.2)
Radiation Therapy no. (%)	278 (45)
NOS no. (%)	17 (2.8)
Vital status	
Dead no. (%)	427 (69.2)
Alive no. (%)	151 (24.5)
NOS no. (%)	39 (6.3)
Average FU (months)	14.7
PTEN Mutations no. pts (%)	137 (22.2)

FU: Follow-up; IDH: Isocitric Dehydrogenase; no.: Number; NOS: Not Otherwise Specified; PTEN: phosphatase and tensin homolog. SD: standard deviation; yrs: years.

PTEN was the most frequent gene by the transcriptome examination (22.2%). It was followed by TTN (20.75%), TP53 (20.10%), EGFR (17.18%), FLG (12.64%), MUC16 (11.51%), NF1 (8.27%), RYR2 (7.62%), PKHD1 (7.29%), HMCN1 (7.29%), SYNE1 (7.29%), SPTA1 (6.97%), PIK3R1 (6.97%), RB1 (6.81%), ATRX (6.65%), IDH (6.62%), PIK3CA (6.48%), OBSCN (6.48%), APOB (6.32%), FLG2 (6.32%), and LRP2 (6.16%) (Figure 1).

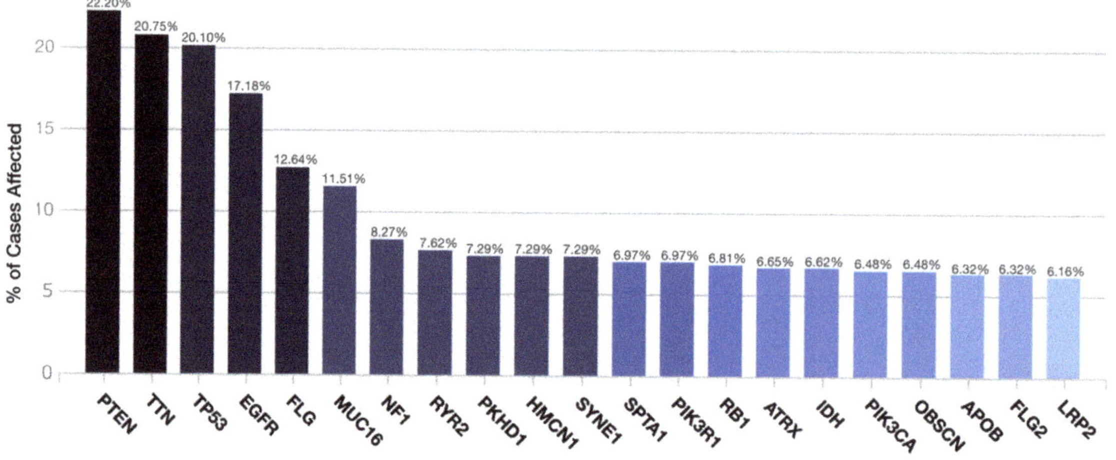

Figure 1. Distribution of the most frequent mutated genes in the TCGA-GBM project.

Expression profiles and mutations of the top 50 genes in TCGA-GBM are shown in the Oncogrid (Figure 2).

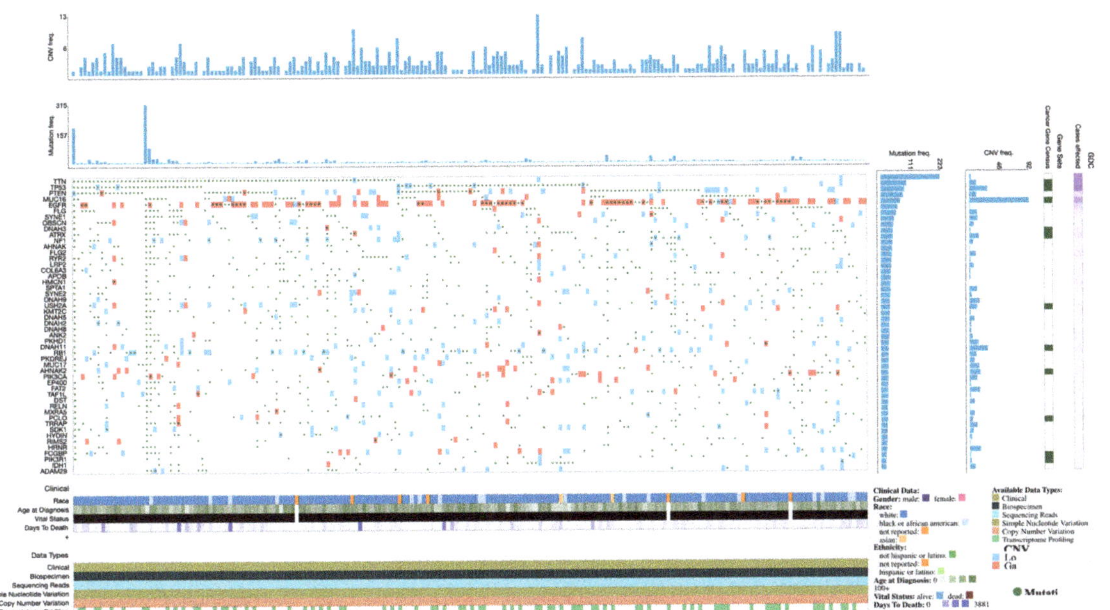

Figure 2. OncoGrid of top 50 mutated genes with impact mutations on the TCGC-GBM cohort.

From the GTEx dataset, a total of 255 samples of normal brain tissue were included. The differential analysis revealed a higher expression of PTEN mRNA levels in the tumor than in normal tissue, albeit not significant, with a log2-transcriptome per million reads (TPM) +1 of 2.8–5.75 and 2.7–4.8 in the TCGA-GBM and GTEx datasets, respectively (Figure 3).

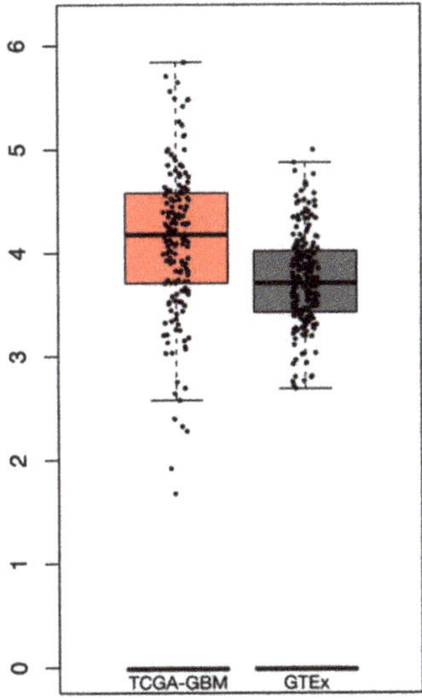

Figure 3. Box plots revealing the differential PTEN mRNA expression levels in TCGA-GBM and GTEx datasets.

3.2. Survival Analysis

Kaplan–Meier curves showed a worse, though non statistic, OS (Log-rank $p = 0.58$, HR = 0.91, $p = 0.6$) and DFS (Log-rank $p = 0.78$, HR = 1, $p = 0.82$) in high-PTEN mutation profile (Figure 4).

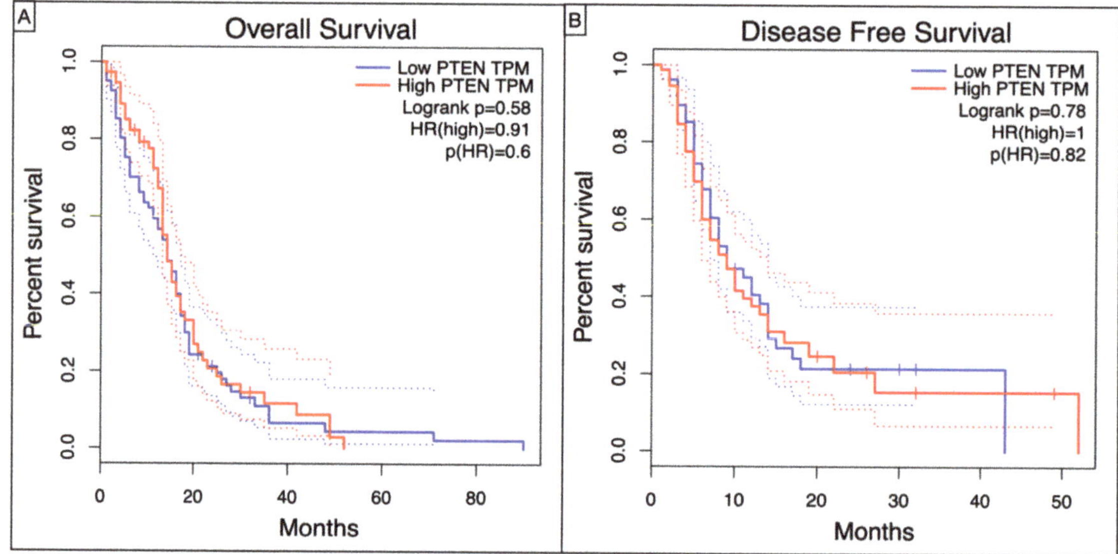

Figure 4. Kaplan–Meyer curves showing the (**A**) Overall Survival and (**B**) Disease Free Survival in TCGA-GBM patients according to the level of PTEN TPM.

3.3. Immune Landscape in PTEN-Related Glioma Microenvironment

Spearman's correlation analysis of immune subpopulations applied in the TCGA-GBM cohort revealed a negative correlation between mRNA PTEN mutations expression (log2 TPM) and infiltrating T cell CD4+ Th 1 ($\rho = -0.245$; $p = 3.81 \times 10^{-3}$), Th 2 ($\rho = -0.225$, $p = 8.06 \times 10^{-3}$), and NK cells ($\rho = -0.163$, $p = 5.64 \times 10^{-2}$) within the tumor microenvironment. Conversely, T_{reg} ($\rho = 0.179$, $p = 3.54 \times 10^{-2}$), endothelial cells ($\rho = 0.303$, $p = 2.97 \times 10^{-4}$), and monocyte/macrophages ($\rho = 0.368$, $p = 9.17 \times 10^{-6}$) were predominant, with a polarization of M2 ($\rho = 0.303$, $p = 2.97 \times 10^{-4}$) against of the monocyte ($\rho = -0.205$, $p = 1.6 \times 10^{-2}$) (Figures 5 and 6).

The GEPIA analysis of the mRNA PTEN mutations expression (logTPM +1) in the immune infiltrates of the TCGA-GBM cohort highlighted the decreased density of B cell naive ($p < 1 \times 10^{-15}$), T cell CD4+ naive ($p < 1 \times 10^{-15}$), T cell CD8+ ($p = 0.08$), and NK cell ($p = 3.97 \times 10^{-9}$) in the glioma microenvironment compared to GTEx samples. The T_{reg} subtype was slightly more represented ($p = 0.07$).

Furthermore, the boxplots evidenced the monocytes ($p < 1 \times 10^{-15}$), endothelial cells ($p < 1 \times 10^{-15}$), macrophages M0 ($p = 3.49 \times 10^{-3}$), M1 ($p < 1 \times 10^{-15}$), and M2 ($p < 1 \times 10^{-15}$) as the main components of the immune profile at the tumor site in the TCGA-GBM cohort (Figures 7 and 8).

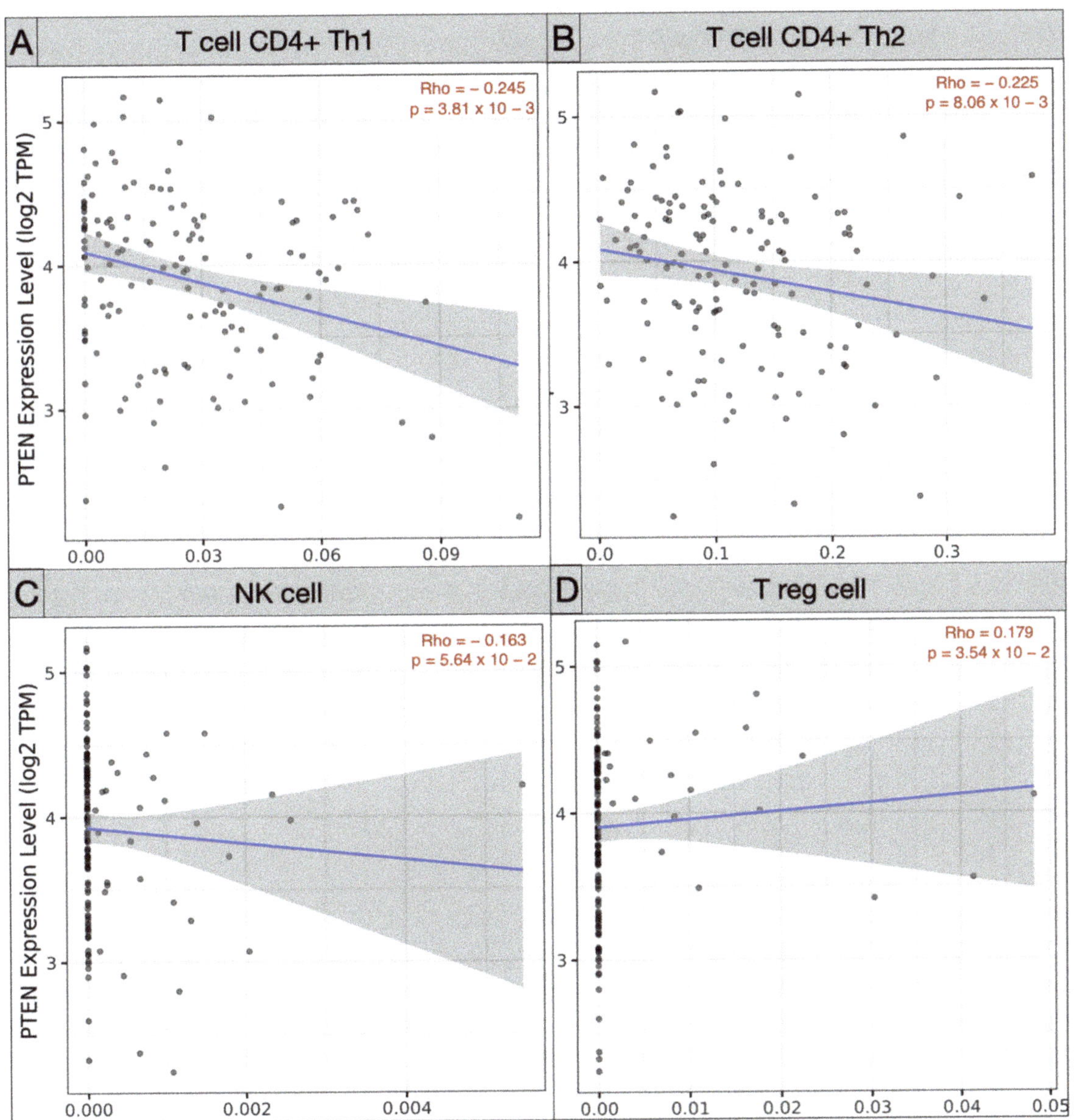

Figure 5. Scatter plots picturing the correlation of mRNA PTEN expression and the immune infiltration of T cell CD4 Th1 (**A**), Th2 (**B**), NK cell (**C**), and T_{reg} (**D**) in the TCGC-GBM project.

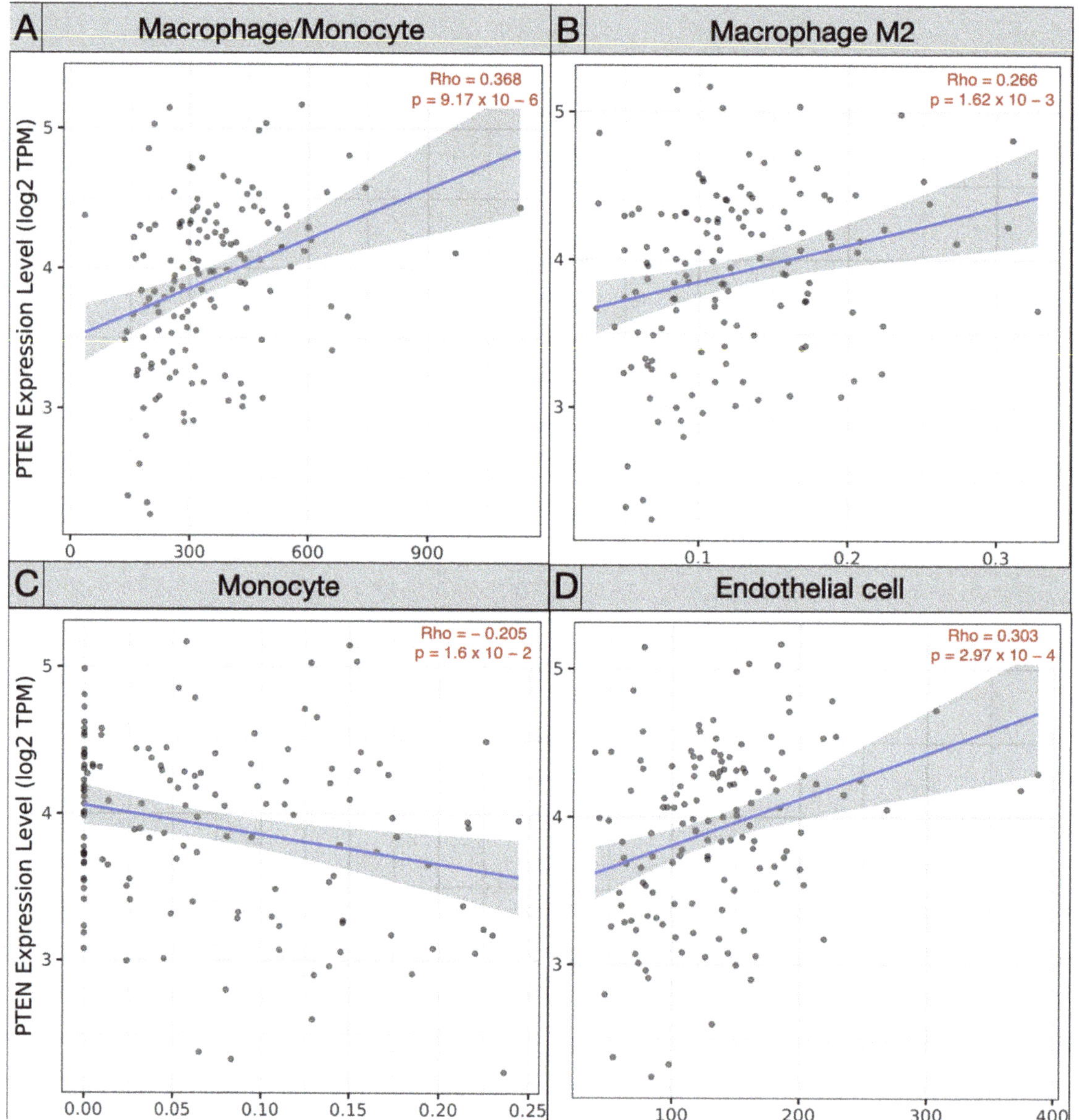

Figure 6. Scatter plots exhibiting the correlation of mRNA PTEN expression and the immune infiltration of (**A**) macrophage/monocyte, (**B**) macrophage M2, (**C**) monocyte, and (**D**) endothelial cells in the TCGA-GBM project.

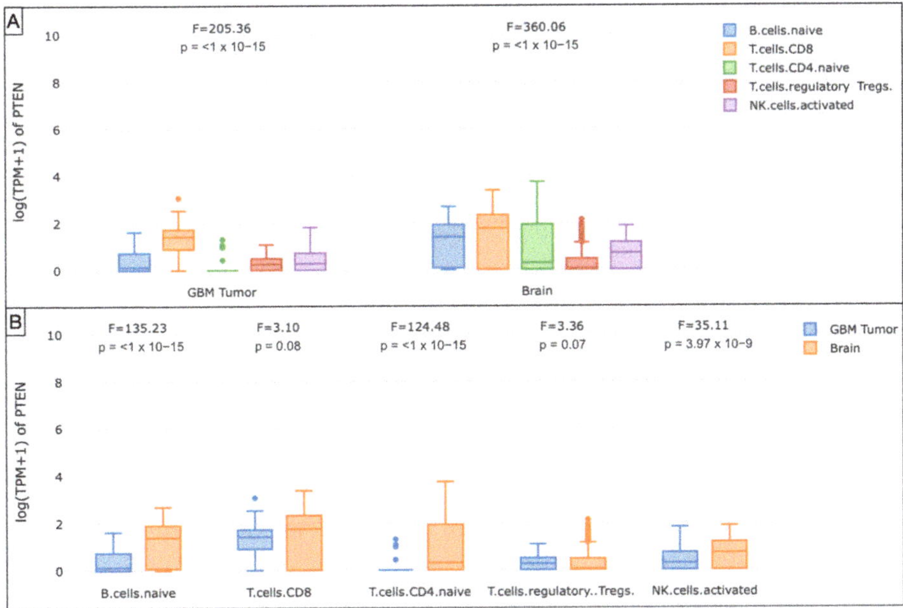

Figure 7. Box plots showing the immune cell subexpression analysis in PTEN-mutated TCGA-GBMs and normal brain samples from GTEx. (**A**) Grouped by tissue. (**B**) Grouped by B cells naive, T cells CD8, CD4, T_{reg}, and NK cells.

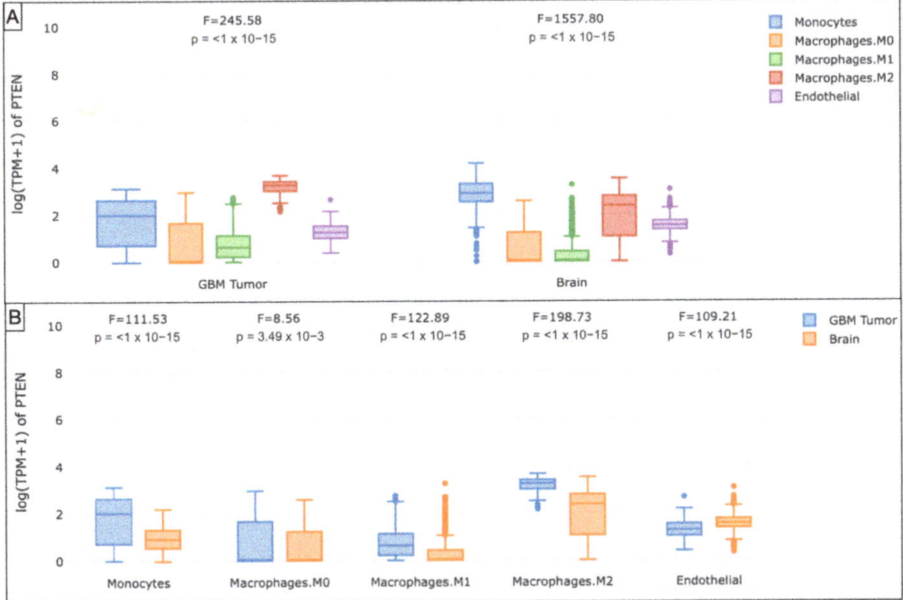

Figure 8. Box plots illustrating the immune cell subexpression analysis in PTEN-mutated TCGA-GBMs and normal brain samples from GTEx. (**A**) Grouped by tissue. (**B**) Grouped by monocytes, macrophages M0, M1, and M2, and endothelial cells.

3.4. Prediction of Chemotherapeutic Response

Based on the GDSC pharmacogenomic database, lower differential targeted responses to TMZ were found in PTEN-mutated and PTEN wild-type samples, with a median IC50 of 531.13 and 701.55 μM, respectively (Figure 9).

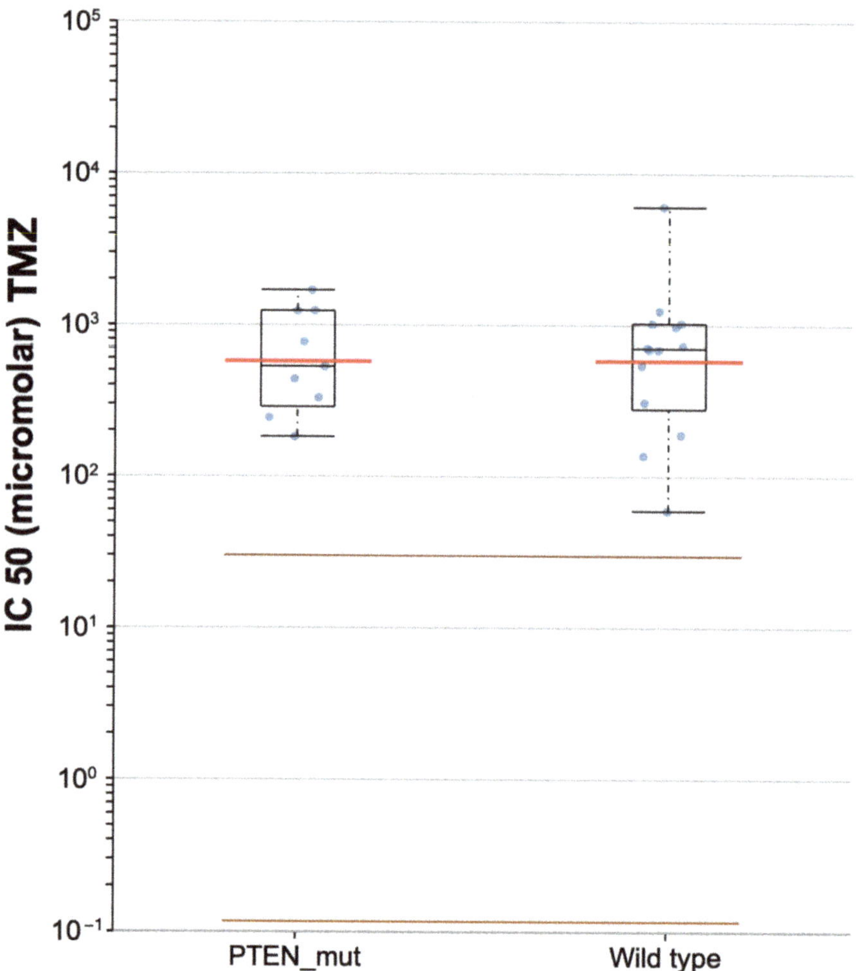

Figure 9. Scatter plot for the sensitivity to temozolomide (TMZ) in PTEN mutated and wild-type of TCGA-GBMs.

4. Discussion

The present study analyzes mutation profiles and immune signatures of PTEN-associated microenvironment estimating GBM patients' prognosis, survival, and chemotherapy response. The PTEN gene regulates the cell cycle and DNA repair mechanisms. Its expression modulates cell proliferation, neural development, and gliogenesis [21,27,42].

PTEN mutations are hallmarks of glioma malignancy and influence the patients' survival [43–49]. As recently reported by Erira and his group, alterations of PTEN genes are related to glioma proliferation. PTEN mutation may induce post-translational changes in low-grade gliomas, leading to malignant progression [50].

The prognostic role of the PTEN status has been widely deepened in the literature. In 2001, Sasaki et al. analyzed the different median survival of glioma patients as it related to the PTEN expression. They reported a better OS for wild-type PTEN gliomas (123.4 months), compared to the mutated ones (14.8 months) [51]. In 2016, Han and colleagues explored the genetic linkage between PTEN expression and patients' outcomes in a meta-analysis where a worse prognosis was revealed for PTEN-mutated gliomas [49]. Zhang and his group, 2021, conducted an online bioinformatics analysis about PTEN mutation as a prognostic signature for HGGs. They designed a tailored risk score based on the individual PTEN status, aiming to simplify HGGs diagnosis, prognosis, and treatment planning. They identified 14 independent prognostic genes in PTENwild-type tumors and 3 for the PTEN-mutated ones. These last proved to be related to the worse survival [52].

Even without any statistical significance, our Kaplan–Meier analysis confirmed dismal OS (Log-rank $p = 0.58$, HR $= 0.91$, $p = 0.6$) and DFS (Log-rank $p = 0.78$, HR $= 1$, $p = 0.82$) for the high-PTEN mutation profiles.

Amid the PTEN-related genetic mechanisms underlying the tumorigenesis, the maintenance of the glioma immune microenvironment is critical. Based on this rationale, our analysis also aimed at typifying the subpopulations in the glioma immune niche.

Our Spearman's correlation tests of immune subpopulations reported a negative correlation ($\rho < 0$) between PTEN mutations and the expression of infiltrating T cell CD4+ Th 1 ($\rho = -0.245$), Th 2 ($\rho = -0.225$), and NK cells ($\rho = -0.163$) within the tumor microenvironment. The GEPIA one-way ANOVA analysis also revealed a lower density of B cell naive ($p < 1 \times 10^{-15}$), CD8+ ($p = 0.08$), CD4+ naive ($p < 1 \times 10^{-15}$), and NK cell ($p = 3.97 \times 10^{-9}$) concomitant with TCGA-GBM high-PTEN mutation profiles, compared to the GTEx samples.

In accordance with the evidence in the literature, these data denoted that brain tumor growth and progression are sustained by genetic mechanisms of immune tolerance and exhaustion. The suppression of T, B, and NK cells activity within the PTEN tumor microenvironment suggested immune-mediated biological processes are involved in pathways for glioma immune evasion and resistance to chemotherapies [53–56].

On the contrary, through Spearman's analysis the T_{reg} ($\rho = 0.179$), endothelial cells ($\rho = 0.303$), and monocyte/macrophages ($\rho = 0.368$) were found to be prevalent. The differential examination demonstrated the polarization of M2 ($\rho = 0.303$) within the glioma microenvironment, versus the monocytes ($\rho = -0.205$). The GEPIA T_{reg} ($p = 0.07$) and macrophages ($p = 3.49 \times 10^{-3}$) were the most represented subtypes, with a prevalence of M2 in the TCGA-GBM group.

T_{regs} preserve immune homeostasis, contribute to the downregulation of T cell activity, and regulate innate and adaptive responses against self-antigens, allergens, and infectious agents [57–61]. T_{regs} also act as immunosuppressive within the tumor microenvironment, repressing the function of CD4+, CD8+, and NK cells [62]. The immune control is carried out by cytokines, extracellular vesicles, perforins, and cytolytic enzymes [63]. They repress antitumor immunity and facilitate immune escape mechanisms, resulting in glioma progression and relapse [64–66].

Our results also verified that the immunosuppressive anticancer microenvironment is sustained by the recruitment of monocytes, which in the glioma context are converted into macrophages, with an explicit M2 polarization. M2 macrophages are known to hold an immunosuppressive role [67,68]. M2 phenotypes induce the differential expression of receptors, cytokines, and chemokines. They produce IL-10, IL-1, and IL-6, stimulating tumorigenesis and negatively affecting the prognosis [69,70]. The M2 macrophages, detected in perivascular areas, enhance the VEGF and COX2 production resulting in increased and aberrant angiogenesis [71–73].

Therefore, T_{reg} and M2 cells stimulate the glioma cell proliferation, invasion, and support immune escape mechanisms [74–77]. These data were confirmed by our bioinformatic analysis ($\rho = 0.3.03$).

Opposing our results, the latest study by Zhou and colleagues published in 2022, found the high expression of PTEN related to a better prognosis for HGGs patients [78]. They discovered via the transwell and flow cytometry that the PTEN gene may inhibit the M1/M2 polarization and M2 macrophages recruitment. These data suggest a potential positive role of the PTEN as an antitumoral immunoregulatory gene [78].

The discrepancy in our study can be explained by the distinct patients cohorts involved, such as the Chinese Glioma Genome Atlas (CGGA) database, and the different techniques applied for data analysis. However, above all, they explored the effects of different PTEN statuses, split into PTEN deletion, PTEN mutated, and PTEN wild-type. This distinction allowed us to assess the specific impact of each PTEN gene expression on glioma immunity. The study demonstrates that, despite advances in genomics, further research is needed to shed light on PTEN activity and its immunological role in tumor progression.

The composition of immune infiltrates explains the prognosis concomitant to the PTEN status and, above all, the ineffective response to standard therapies. TMZ is currently the first line of treatment for HGGs in combination with surgery and radiotherapy [79,80]. It is still debated whether the PTEN mutation may influence sensitivity to radiochemotherapy [81–83]. PTEN controls the Wnt/β-catenin and PI3K/Akt/mTOR signaling pathways and arrests the cell cycle at the G2/M phase. TMZ alkylates DNA at this stage, hence PTEN overexpression affects the complex biochemical mechanism of drug alkylation encouraging TMZ activity [84,85].

In 2012 Carico and colleagues conducted a clinical study involving newly diagnosed GBM treated with TMZ. They reported greater effectiveness of TMZ in GBMs with PTEN loss. Inaba et al. investigated the effects of TMZ related to PTEN status founding an increased efficacy in cases of PTEN mutation [83]. Similarly, our analysis of the GDSC database revealed a lower IC50 for PTEN-mutated GBMs (531.13 μM) in comparison with wild-types (701.55 μM).

Apart from conventional chemotherapy, the identification of immune phenotypes and molecular interactions within the tumor microenvironment has been recognized as crucial to widening the spectrum of tailored strategies against the immune escape mechanisms [86–88].

Limitations of the Study

The present study has several undeniable limitations, among which are the relatively limited number of patients and short follow-up (average 14.7 months). Other potential biases were the different patients' ethnicity, limited data about the histological classification, and heterogeneity of radiochemotherapy regimens.

5. Conclusions

PTEN mutations frequently occur in malignant brain tumors, contributing to their progression, reduced OS, and DFS. Within the glioblastoma microenvironment, the PTEN-related immune landscape mainly consists of T_{reg} and M2 macrophages. They repress the antitumor immune activation and are responsible for the triggering of the glioma cell growth, invasion, and aberrant vasculogenesis.

PTEN expression and related glioma microenvironment also influence the sensitivity to conventional radiochemotherapy.

Prospective and randomized trials are necessary to validate these data and to develop novel target treatments.

Author Contributions: Writing—Original Draft, Conceptualization, Writing—Original Draft, A.G.L.; Writing—Review & Editing, Supervision, Validation, S.L. All authors have read and agreed to the published version of the manuscript.

Funding: This study was supported by Rete Italiana delle Neuroscienze with the Network Research Project (#RCR-2019-23669119-002).

Institutional Review Board Statement: All procedures performed in the study were in accordance with the ethical standards of the institution and with the 1964 Helsinki Declaration and its later amendments or comparable ethical standards.

Data Availability Statement: All data are included in the main text.

Conflicts of Interest: The authors declare no conflict of interest.

References

1. Ostrom, Q.T.; Gittleman, H.; Liao, P.; Rouse, C.; Chen, Y.; Dowling, J.; Wolinsky, Y.; Kruchko, C.; Barnholtz-Sloan, J. CBTRUS statistical report: Primary brain and central nervous system tumors diagnosed in the United States in 2007–2011. *Neuro Oncol.* **2014**, *16* (Suppl. S4), iv1–iv63. [CrossRef] [PubMed]
2. Ferlay, J.; Steliarova-Foucher, E.; Lortet-Tieulent, J.; Rosso, S.; Coebergh, J.W.; Comber, H.; Forman, D.; Bray, F. Cancer incidence and mortality patterns in Europe: Estimates for 40 countries in 2012. *Eur. J. Cancer* **2013**, *49*, 1374–1403. [CrossRef] [PubMed]
3. Soerjomataram, I.; Lortet-Tieulent, J.; Parkin, D.M.; Ferlay, J.; Mathers, C.; Forman, D.; Bray, F. Global burden of cancer in 2008: A systematic analysis of disability-adjusted life-years in 12 world regions. *Lancet* **2012**, *380*, 1840–1850. [CrossRef]
4. Ladomersky, E.; Genet, M.; Zhai, L.; Gritsina, G.; Lauing, K.L.; Lulla, R.R.; Fangusaro, J.; Lenzen, A.; Kumthekar, P.; Raizer, J.J.; et al. Improving vaccine efficacy against malignant glioma. *Oncoimmunology* **2016**, *5*, e1196311. [CrossRef]
5. Ostrom, Q.T.; Gittleman, H.; Stetson, L.; Virk, S.M.; Barnholtz-Sloan, J.S. Epidemiology of gliomas. *Cancer Treat. Res.* **2015**, *163*, 1–14. [CrossRef]
6. Wrensch, M.; Minn, Y.; Chew, T.; Bondy, M.; Berger, M.S. Epidemiology of primary brain tumors: Current concepts and review of the literature. *Neuro Oncol.* **2002**, *4*, 278–299. [CrossRef]
7. Leenstra, S.; Oskam, N.T.; Bijleveld, E.H.; Bosch, D.A.; Troost, D.; Hulsebos, T.J. Genetic sub-types of human malignant astrocytoma correlate with survival. *Int. J. Cancer* **1998**, *79*, 159–165. [CrossRef]
8. Attarian, F.; Taghizadeh-Hesary, F.; Fanipakdel, A.; Javadinia, S.A.; Porouhan, P.; PeyroShabany, B.; Fazilat-Panah, D. A Systematic Review and Meta-Analysis on the Number of Adjuvant Temozolomide Cycles in Newly Diagnosed Glioblastoma. *Front. Oncol.* **2021**, *11*, 1–6. [CrossRef]
9. Johnson, D.R.; O'Neill, B.P. Glioblastoma survival in the United States before and during the temozolomide era. *J. Neurooncol.* **2012**, *107*, 359–364. [CrossRef]
10. Eagan, R.T.; Scott, M. Evaluation of prognostic factors in chemotherapy of recurrent brain tumors. *J. Clin. Oncol.* **1983**, *1*, 38–44. [CrossRef]
11. Wen, P.Y.; Kesari, S. Malignant gliomas in adults. *N. Engl. J. Med.* **2008**, *359*, 492–507. [CrossRef]
12. Perry, J.R.; Laperriere, N.; O'Callaghan, C.J.; Brandes, A.A.; Menten, J.; Phillips, C.; Fay, M.; Nishikawa, R.; Cairncross, J.G.; Roa, W.; et al. Short-Course Radiation plus Temozolomide in Elderly Patients with Glioblastoma. *N. Engl. J. Med.* **2017**, *376*, 1027–1037. [CrossRef]
13. Cancer Genome Atlas (TCGA) Research Network. Comprehensive genomic characterization defines human glioblastoma genes and core pathways. *Nature* **2008**, *455*, 1061–1068. [CrossRef]
14. Parsons, D.W.; Jones, S.; Zhang, X.; Lin, J.C.; Leary, R.J.; Angenendt, P.; Mankoo, P.; Carter, H.; Siu, I.M.; Gallia, G.L.; et al. An integrated genomic analysis of human glioblastoma multiforme. *Science* **2008**, *321*, 1807–1812. [CrossRef]
15. Verhaak, R.G.; Hoadley, K.A.; Purdom, E.; Wang, V.; Qi, Y.; Wilkerson, M.D.; Miller, C.R.; Ding, L.; Golub, T.; Mesirov, J.P.; et al. Integrated genomic analysis identifies clinically relevant subtypes of glioblastoma characterized by abnormalities in PDGFRA, IDH1, EGFR, and NF1. *Cancer Cell* **2010**, *17*, 98–110. [CrossRef]
16. Palumbo, P.; Lombardi, F.; Siragusa, G.; Dehcordi, S.R.; Luzzi, S.; Cimini, A.; Cifone, M.G.; Cinque, B. Involvement of NOS2 Activity on Human Glioma Cell Growth, Clonogenic Potential, and Neurosphere Generation. *Int. J. Mol. Sci.* **2018**, *19*, 2801. [CrossRef]
17. Palumbo, P.; Lombardi, F.; Augello, F.R.; Giusti, I.; Luzzi, S.; Dolo, V.; Cifone, M.G.; Cinque, B. NOS_2 inhibitor 1400W Induces Autophagic Flux and Influences Extracellular Vesicle Profile in Human Glioblastoma U87MG Cell Line. *Int. J. Mol. Sci.* **2019**, *20*, 3010. [CrossRef]
18. Louis, D.N.; Perry, A.; Wesseling, P.; Brat, D.J.; Cree, I.A.; Figarella-Branger, D.; Hawkins, C.; Ng, H.K.; Pfister, S.M.; Reifenberger, G.; et al. The 2021 WHO Classification of Tumors of the Central Nervous System: A summary. *Neuro Oncol.* **2021**, *23*, 1231–1251. [CrossRef]
19. James, C.D.; Galanis, E.; Frederick, L.; Kimmel, D.W.; Cunningham, J.M.; Atherton-Skaff, P.J.; O'Fallon, J.R.; Jenkins, R.B.; Buckner, J.C.; Hunter, S.B.; et al. Tumor suppressor gene alterations in malignant gliomas: Histopathological associations and prognostic evaluation. *Int. J. Oncol.* **1999**, *15*, 547–553. [CrossRef]
20. Rasheed, B.K.; Stenzel, T.T.; McLendon, R.E.; Parsons, R.; Friedman, A.H.; Friedman, H.S.; Bigner, D.D.; Bigner, S.H. PTEN gene mutations are seen in high-grade but not in low-grade gliomas. *Cancer Res.* **1997**, *57*, 4187–4190.
21. Yang, Y.; Shao, N.; Luo, G.; Li, L.; Zheng, L.; Nilsson-Ehle, P.; Xu, N. Mutations of PTEN gene in gliomas correlate to tumor differentiation and short-term survival rate. *Anticancer Res.* **2010**, *30*, 981–985. [PubMed]

22. Javadinia, S.A.; Shahidsales, S.; Fanipakdel, A.; Joudi-Mashhad, M.; Mehramiz, M.; Talebian, S.; Maftouh, M.; Mardani, R.; Hassanian, S.M.; Khazaei, M.; et al. Therapeutic potential of targeting the Wnt/β-catenin pathway in the treatment of pancreatic cancer. *J. Cell. Biochem.* **2018**, *120*, 6833–6840. [CrossRef] [PubMed]

23. Sun, H.; Lesche, R.; Li, D.M.; Liliental, J.; Zhang, H.; Gao, J.; Gavrilova, N.; Mueller, B.; Liu, X.; Wu, H. PTEN modulates cell cycle progression and cell survival by regulating phosphatidylinositol 3,4,5,-trisphosphate and Akt/protein kinase B signaling pathway. *Proc. Natl. Acad. Sci. USA* **1999**, *96*, 6199–6204. [CrossRef] [PubMed]

24. Li, J.; Yen, C.; Liaw, D.; Podsypanina, K.; Bose, S.; Wang, S.I.; Puc, J.; Miliaresis, C.; Rodgers, L.; McCombie, R.; et al. PTEN, a putative protein tyrosine phosphatase gene mutated in human brain, breast, and prostate cancer. *Science* **1997**, *275*, 1943–1947. [CrossRef]

25. Cantley, L.C. The phosphoinositide 3-kinase pathway. *Science* **2002**, *296*, 1655–1657. [CrossRef]

26. Javadinia, S.A.; Shahidsales, S.; Fanipakdel, A.; Mostafapour, A.; Joudi-Mashhad, M.; Ferns, G.A.; Avan, A. The Esophageal Cancer and the PI3K/AKT/mTOR Signaling Regulatory microRNAs: A Novel Marker for Prognosis, and a Possible Target for Immunotherapy. *Curr. Pharm. Des.* **2018**, *24*, 4646–4651. [CrossRef]

27. Ming, M.; He, Y.Y. PTEN in DNA damage repair. *Cancer Lett.* **2012**, *319*, 125–129. [CrossRef]

28. Yang, Y.; Shao, N.; Luo, G.; Li, L.; Nilsson-Ehle, P.; Xu, N. Relationship between PTEN gene expression and differentiation of human glioma. *Scand. J. Clin. Lab. Investig.* **2006**, *66*, 469–475. [CrossRef]

29. Kitange, G.J.; Templeton, K.L.; Jenkins, R.B. Recent advances in the molecular genetics of primary gliomas. *Curr. Opin. Oncol.* **2003**, *15*, 197–203. [CrossRef]

30. Qin, Z.; Zhang, X.; Chen, Z.; Liu, N. Establishment and validation of an immune-based prognostic score model in glioblastoma. *Int. Immunopharmacol.* **2020**, *85*, 106636. [CrossRef]

31. Gieryng, A.; Pszczolkowska, D.; Walentynowicz, K.A.; Rajan, W.D.; Kaminska, B. Immune microenvironment of gliomas. *Lab. Investig.* **2017**, *97*, 498–518. [CrossRef]

32. Wilcox, J.A.; Ramakrishna, R.; Magge, R. Immunotherapy in Glioblastoma. *World Neurosurg.* **2018**, *116*, 518–528. [CrossRef]

33. Grabowski, M.M.; Sankey, E.W.; Ryan, K.J.; Chongsathidkiet, P.; Lorrey, S.J.; Wilkinson, D.S.; Fecci, P.E. Immune suppression in gliomas. *J. Neurooncol.* **2021**, *151*, 3–12. [CrossRef]

34. Hanaei, S.; Afshari, K.; Hirbod-Mobarakeh, A.; Mohajer, B.; Amir Dastmalchi, D.; Rezaei, N. Therapeutic efficacy of specific immunotherapy for glioma: A systematic review and meta-analysis. *Rev. Neurosci.* **2018**, *29*, 443–461. [CrossRef]

35. Giotta Lucifero, A.; Luzzi, S.; Brambilla, I.; Trabatti, C.; Mosconi, M.; Savasta, S.; Foiadelli, T. Innovative therapies for malignant brain tumors: The road to a tailored cure. *Acta Biomed.* **2020**, *91*, 5–17. [CrossRef]

36. Luzzi, S.; Crovace, A.M.; Del Maestro, M.; Giotta Lucifero, A.; Elbabaa, S.K.; Cinque, B.; Palumbo, P.; Lombardi, F.; Cimini, A.; Cifone, M.G.; et al. The cell-based approach in neurosurgery: Ongoing trends and future perspectives. *Heliyon* **2019**, *5*, e02818. [CrossRef]

37. Giotta Lucifero, A.; Luzzi, S. Against the Resilience of High-Grade Gliomas: The Immunotherapeutic Approach (Part I). *Brain Sci.* **2021**, *11*, 386. [CrossRef]

38. Giotta Lucifero, A.; Luzzi, S. Against the Resilience of High-Grade Gliomas: Gene Therapies (Part II). *Brain Sci.* **2021**, *11*, 976. [CrossRef]

39. Luzzi, S.; Giotta Lucifero, A.; Brambilla, I.; Trabatti, C.; Mosconi, M.; Savasta, S.; Foiadelli, T. The impact of stem cells in neuro-oncology: Applications, evidence, limitations and challenges. *Acta Biomed.* **2020**, *91*, 51–60. [CrossRef]

40. Colaprico, A.; Silva, T.C.; Olsen, C.; Garofano, L.; Cava, C.; Garolini, D.; Sabedot, T.S.; Malta, T.M.; Pagnotta, S.M.; Castiglioni, I.; et al. TCGAbiolinks: An R/Bioconductor package for integrative analysis of TCGA data. *Nucleic Acids Res.* **2016**, *44*, e71. [CrossRef]

41. Yang, W.; Soares, J.; Greninger, P.; Edelman, E.J.; Lightfoot, H.; Forbes, S.; Bindal, N.; Beare, D.; Smith, J.A.; Thompson, I.R.; et al. Genomics of Drug Sensitivity in Cancer (GDSC): A resource for therapeutic biomarker discovery in cancer cells. *Nucleic Acids Res.* **2012**, *41*, D955–D961. [CrossRef] [PubMed]

42. Ortega-Molina, A.; Serrano, M. PTEN in cancer, metabolism, and aging. *Trends Endocrinol. Metab.* **2013**, *24*, 184–189. [CrossRef] [PubMed]

43. Smith, J.S.; Tachibana, I.; Passe, S.M.; Huntley, B.K.; Borell, T.J.; Iturria, N.; O'Fallon, J.R.; Schaefer, P.L.; Scheithauer, B.W.; James, C.D.; et al. PTEN mutation, EGFR amplification, and outcome in patients with anaplastic astrocytoma and glioblastoma multiforme. *J. Natl. Cancer Inst.* **2001**, *93*, 1246–1256. [CrossRef] [PubMed]

44. Srividya, M.R.; Thota, B.; Shailaja, B.C.; Arivazhagan, A.; Thennarasu, K.; Chandramouli, B.A.; Hegde, A.S.; Santosh, V. Homozygous 10q23/PTEN deletion and its impact on outcome in glioblastoma: A prospective translational study on a uniformly treated cohort of adult patients. *Neuropathology* **2011**, *31*, 376–383. [CrossRef] [PubMed]

45. Koul, D. PTEN signaling pathways in glioblastoma. *Cancer Biol. Ther.* **2008**, *7*, 1321–1325. [CrossRef] [PubMed]

46. D'Urso, O.F.; D'Urso, P.I.; Marsigliante, S.; Storelli, C.; Luzi, G.; Gianfreda, C.D.; Montinaro, A.; Distante, A.; Ciappetta, P. Correlative analysis of gene expression profile and prognosis in patients with gliomatosis cerebri. *Cancer* **2009**, *115*, 3749–3757. [CrossRef] [PubMed]

47. Abdullah, J.M.; Farizan, A.; Asmarina, K.; Zainuddin, N.; Ghazali, M.M.; Jaafar, H.; Isa, M.N.; Naing, N.N. Association of loss of heterozygosity and PTEN gene abnormalities with paraclinical, clinical modalities and survival time of glioma patients in Malaysia. *Asian J. Surg.* **2006**, *29*, 274–282. [CrossRef]

48. Xu, J.; Li, Z.; Wang, J.; Chen, H.; Fang, J.Y. Combined PTEN Mutation and Protein Expression Associate with Overall and Disease-Free Survival of Glioblastoma Patients. *Transl. Oncol.* **2014**, *7*, 196–205.e191. [CrossRef]
49. Han, F.; Hu, R.; Yang, H.; Liu, J.; Sui, J.; Xiang, X.; Wang, F.; Chu, L.; Song, S. PTEN gene mutations correlate to poor prognosis in glioma patients: A meta-analysis. *Onco Targets Ther.* **2016**, *9*, 3485–3492. [CrossRef]
50. Erira, A.; Velandia, F.; Penagos, J.; Zubieta, C.; Arboleda, G. Differential Regulation of the EGFR/PI3K/AKT/PTEN Pathway between Low- and High-Grade Gliomas. *Brain Sci.* **2021**, *11*, 1655. [CrossRef]
51. Sasaki, H.; Zlatescu, M.C.; Betensky, R.A.; Ino, Y.; Cairncross, J.G.; Louis, D.N. PTEN is a target of chromosome 10q loss in anaplastic oligodendrogliomas and PTEN alterations are associated with poor prognosis. *Am. J. Pathol.* **2001**, *159*, 359–367. [CrossRef]
52. Zhang, P.; Meng, X.; Liu, L.; Li, S.; Li, Y.; Ali, S.; Li, S.; Xiong, J.; Liu, X.; Li, S.; et al. Identification of the Prognostic Signatures of Glioma With Different PTEN Status. *Front. Oncol.* **2021**, *11*, 633357. [CrossRef]
53. Cordell, E.C.; Alghamri, M.S.; Castro, M.G.; Gutmann, D.H. T lymphocytes as dynamic regulators of glioma pathobiology. *Neuro Oncol.* **2022**, noac055. [CrossRef]
54. Codrici, E.; Popescu, I.D.; Tanase, C.; Enciu, A.M. Friends with Benefits: Chemokines, Glioblastoma-Associated Microglia/Macrophages, and Tumor Microenvironment. *Int. J. Mol. Sci.* **2022**, *23*, 2509. [CrossRef]
55. DeCordova, S.; Shastri, A.; Tsolaki, A.G.; Yasmin, H.; Klein, L.; Singh, S.K.; Kishore, U. Molecular Heterogeneity and Immunosuppressive Microenvironment in Glioblastoma. *Front. Immunol.* **2020**, *11*, 1402. [CrossRef]
56. Ma, Q.; Long, W.; Xing, C.; Chu, J.; Luo, M.; Wang, H.Y.; Liu, Q.; Wang, R.F. Cancer Stem Cells and Immunosuppressive Microenvironment in Glioma. *Front. Immunol.* **2018**, *9*, 2924. [CrossRef]
57. Grover, P.; Goel, P.N.; Greene, M.I. Regulatory T Cells: Regulation of Identity and Function. *Front. Immunol.* **2021**, *12*, 750542. [CrossRef]
58. Strauss, L.; Bergmann, C.; Whiteside, T.L. Human circulating CD4+CD25highFoxp3+ regulatory T cells kill autologous CD8+ but not CD4+ responder cells by Fas-mediated apoptosis. *J. Immunol.* **2009**, *182*, 1469–1480. [CrossRef]
59. Boer, M.C.; Joosten, S.A.; Ottenhoff, T.H. Regulatory T-Cells at the Interface between Human Host and Pathogens in Infectious Diseases and Vaccination. *Front. Immunol.* **2015**, *6*, 217. [CrossRef]
60. Sakaguchi, S. Regulatory T cells: Key controllers of immunologic self-tolerance. *Cell* **2000**, *101*, 455–458. [CrossRef]
61. Scheinecker, C.; Göschl, L.; Bonelli, M. Treg cells in health and autoimmune diseases: New insights from single cell analysis. *J. Autoimmun.* **2020**, *110*, 102376. [CrossRef] [PubMed]
62. Togashi, Y.; Shitara, K.; Nishikawa, H. Regulatory T cells in cancer immunosuppression—Implications for anticancer therapy. *Nat. Rev. Clin. Oncol.* **2019**, *16*, 356–371. [CrossRef] [PubMed]
63. Okeke, E.B.; Uzonna, J.E. The Pivotal Role of Regulatory T Cells in the Regulation of Innate Immune Cells. *Front. Immunol.* **2019**, *10*, 680. [CrossRef] [PubMed]
64. Nishikawa, H.; Sakaguchi, S. Regulatory T cells in tumor immunity. *Int. J. Cancer* **2010**, *127*, 759–767. [CrossRef]
65. Takeuchi, Y.; Nishikawa, H. Roles of regulatory T cells in cancer immunity. *Int. Immunol.* **2016**, *28*, 401–409. [CrossRef]
66. Kim, J.H.; Kim, B.S.; Lee, S.K. Regulatory T Cells in Tumor Microenvironment and Approach for Anticancer Immunotherapy. *Immune Netw.* **2020**, *20*, e4. [CrossRef]
67. Gajewski, T.F. Identifying and Overcoming Immune Resistance Mechanisms in the Melanoma Tumor Microenvironment. *Clin. Cancer Res.* **2006**, *12*, 2326s–2330s. [CrossRef]
68. Shevach, E.M. CD4$^+$CD25$^+$ suppressor T cells: More questions than answers. *Nat. Rev. Immunol.* **2002**, *2*, 389–400. [CrossRef]
69. Mantovani, A.; Sozzani, S.; Locati, M.; Allavena, P.; Sica, A. Macrophage polarization: Tumor-associated macrophages as a paradigm for polarized M2 mononuclear phagocytes. *Trends Immunol.* **2002**, *23*, 549–555. [CrossRef]
70. Yunna, C.; Mengru, H.; Lei, W.; Weidong, C. Macrophage M1/M2 polarization. *Eur. J. Pharmacol.* **2020**, *877*, 173090. [CrossRef]
71. Mignogna, C.; Signorelli, F.; Vismara, M.F.; Zeppa, P.; Camastra, C.; Barni, T.; Donato, G.; Di Vito, A. A reappraisal of macrophage polarization in glioblastoma: Histopathological and immunohistochemical findings and review of the literature. *Pathol. Res. Pract.* **2016**, *212*, 491–499. [CrossRef]
72. Hughes, R.; Qian, B.Z.; Rowan, C.; Muthana, M.; Keklikoglou, I.; Olson, O.C.; Tazzyman, S.; Danson, S.; Addison, C.; Clemons, M.; et al. Perivascular M2 Macrophages Stimulate Tumor Relapse after Chemotherapy. *Cancer Res.* **2015**, *75*, 3479–3491. [CrossRef]
73. Fan, Y.; Ye, J.; Shen, F.; Zhu, Y.; Yeghiazarians, Y.; Zhu, W.; Chen, Y.; Lawton, M.T.; Young, W.L.; Yang, G.Y. Interleukin-6 stimulates circulating blood-derived endothelial progenitor cell angiogenesis in vitro. *J. Cereb. Blood Flow Metab.* **2008**, *28*, 90–98. [CrossRef]
74. Chen, Y.; Song, Y.; Du, W.; Gong, L.; Chang, H.; Zou, Z. Tumor-associated macrophages: An accomplice in solid tumor progression. *J. Biomed. Sci.* **2019**, *26*, 78. [CrossRef]
75. Hussain, S.F.; Yang, D.; Suki, D.; Aldape, K.; Grimm, E.; Heimberger, A.B. The role of human glioma-infiltrating microglia/macrophages in mediating antitumor immune responses. *Neuro Oncol.* **2006**, *8*, 261–279. [CrossRef]
76. Zhu, C.; Kros, J.M.; Cheng, C.; Mustafa, D. The contribution of tumor-associated macrophages in glioma neo-angiogenesis and implications for anti-angiogenic strategies. *Neuro Oncol.* **2017**, *19*, 1435–1446. [CrossRef]
77. Qian, B.Z.; Pollard, J.W. Macrophage diversity enhances tumor progression and metastasis. *Cell* **2010**, *141*, 39–51. [CrossRef]
78. Zhou, F.; Shi, Q.; Fan, X.; Yu, R.; Wu, Z.; Wang, B.; Tian, W.; Yu, T.; Pan, M.; You, Y.; et al. Diverse Macrophages Constituted the Glioma Microenvironment and Influenced by PTEN Status. *Front. Immunol.* **2022**, *13*, 841404. [CrossRef]

79. Stupp, R.; Hegi, M.E.; Mason, W.P.; van den Bent, M.J.; Taphoorn, M.J.; Janzer, R.C.; Ludwin, S.K.; Allgeier, A.; Fisher, B.; Belanger, K.; et al. Effects of radiotherapy with concomitant and adjuvant temozolomide versus radiotherapy alone on survival in glioblastoma in a randomised phase III study: 5-year analysis of the EORTC-NCIC trial. *Lancet Oncol.* **2009**, *10*, 459–466. [CrossRef]

80. Stupp, R.; Mason, W.P.; van den Bent, M.J.; Weller, M.; Fisher, B.; Taphoorn, M.J.; Belanger, K.; Brandes, A.A.; Marosi, C.; Bogdahn, U.; et al. Radiotherapy plus concomitant and adjuvant temozolomide for glioblastoma. *N. Engl. J. Med.* **2005**, *352*, 987–996. [CrossRef]

81. Sano, T.; Lin, H.; Chen, X.; Langford, L.A.; Koul, D.; Bondy, M.L.; Hess, K.R.; Myers, J.N.; Hong, Y.K.; Yung, W.K.; et al. Differential expression of MMAC/PTEN in glioblastoma multiforme: Relationship to localization and prognosis. *Cancer Res.* **1999**, *59*, 1820–1824. [PubMed]

82. McEllin, B.; Camacho, C.V.; Mukherjee, B.; Hahm, B.; Tomimatsu, N.; Bachoo, R.M.; Burma, S. PTEN loss compromises homologous recombination repair in astrocytes: Implications for glioblastoma therapy with temozolomide or poly(ADP-ribose) polymerase inhibitors. *Cancer Res.* **2010**, *70*, 5457–5464. [CrossRef] [PubMed]

83. Inaba, N.; Kimura, M.; Fujioka, K.; Ikeda, K.; Somura, H.; Akiyoshi, K.; Inoue, Y.; Nomura, M.; Saito, Y.; Saito, H.; et al. The effect of PTEN on proliferation and drug-, and radiosensitivity in malignant glioma cells. *Anticancer Res.* **2011**, *31*, 1653–1658. [PubMed]

84. Zając, A.; Sumorek-Wiadro, J.; Langner, E.; Wertel, I.; Maciejczyk, A.; Pawlikowska-Pawlęga, B.; Pawelec, J.; Wasiak, M.; Hułas-Stasiak, M.; Bądziul, D.; et al. Involvement of PI3K Pathway in Glioma Cell Resistance to Temozolomide Treatment. *Int. J. Mol. Sci.* **2021**, *22*, 5155. [CrossRef]

85. Carico, C.; Nuño, M.; Mukherjee, D.; Elramsisy, A.; Dantis, J.; Hu, J.; Rudnick, J.; Yu, J.S.; Black, K.L.; Bannykh, S.I.; et al. Loss of PTEN is not associated with poor survival in newly diagnosed glioblastoma patients of the temozolomide era. *PLoS ONE* **2012**, *7*, e33684. [CrossRef]

86. Giotta Lucifero, A.; Luzzi, S.; Brambilla, I.; Schena, L.; Mosconi, M.; Foiadelli, T.; Savasta, S. Potential roads for reaching the summit: An overview on target therapies for high-grade gliomas. *Acta Biomed.* **2020**, *91*, 61–78. [CrossRef]

87. Giotta Lucifero, A.; Luzzi, S.; Brambilla, I.; Guarracino, C.; Mosconi, M.; Foiadelli, T.; Savasta, S. Gene therapies for high-grade gliomas: From the bench to the bedside. *Acta Biomed.* **2020**, *91*, 32–50. [CrossRef]

88. Luzzi, S.; Giotta Lucifero, A.; Brambilla, I.; Magistrali, M.; Mosconi, M.; Savasta, S.; Foiadelli, T. Adoptive immunotherapies in neuro-oncology: Classification, recent advances, and translational challenges. *Acta Biomed.* **2020**, *91*, 18–31. [CrossRef]

Article

Safety and Efficacy of Hypofractionated Stereotactic Radiotherapy with Anlotinib Targeted Therapy for Glioblastoma at the First Recurrence: A Preliminary Report

Yun Guan [1,2,3,4,†], Jing Li [1,2,3,4,†], Xiu Gong [1,2,3,4,†], Huaguang Zhu [1,2,3,4], Chao Li [1,2,3,4], Guanghai Mei [1,2,3,4], Xiaoxia Liu [1,2,3,4], Li Pan [1,2,3,4], Jiazhong Dai [1,2,3,4], Yang Wang [1,2,3,4], Enmin Wang [1,2,3,4], Ying Liu [5,*] and Xin Wang [1,2,3,4,*]

[1]   CyberKnife Center, Department of Neurosurgery, Huashan Hospital, Fudan University, 12 Wulumuqi Road (M), Shanghai 200040, China; yguan10@fudan.edu.cn (Y.G.); lijingck@fudan.edu.cn (J.L.); gongxiu2005@163.com (X.G.); zhuhg@fudan.edu.cn (H.Z.); lichao11@fudan.edu.cn (C.L.); meighai@126.com (G.M.); xiaoxia@fudan.edu.cn (X.L.); lipanmr@sina.com (L.P.); djzhadb@126.com (J.D.); janetcyj@163.com (Y.W.); wangem@fudan.edu.cn (E.W.)
[2]   Neurosurgical Institute, Fudan University, 12 Wulumuqi Road (M), Shanghai 200040, China
[3]   National Center for Neurological Disorders, 12 Wulumuqi Road (M), Shanghai 200040, China
[4]   Shanghai Key Laboratory of Brain Function and Restoration and Neural Regeneration, 12 Wulumuqi Road (M), Shanghai 200040, China
[5]   Department of Pathology, School of Basic Medical Sciences, Fudan University, 138 Yi Xue Yuan Road, Shanghai 200032, China
*   Correspondence: yliu@shmu.edu.cn (Y.L.); wangxinck@fudan.edu.cn (X.W.)
†   These authors contributed equally to this work.

Citation: Guan, Y.; Li, J.; Gong, X.; Zhu, H.; Li, C.; Mei, G.; Liu, X.; Pan, L.; Dai, J.; Wang, Y.; et al. Safety and Efficacy of Hypofractionated Stereotactic Radiotherapy with Anlotinib Targeted Therapy for Glioblastoma at the First Recurrence: A Preliminary Report. *Brain Sci.* **2022**, *12*, 471. https://doi.org/10.3390/brainsci12040471

Academic Editors: N. Scott Litofsky and Stephen D. Meriney

Received: 12 January 2022
Accepted: 24 March 2022
Published: 2 April 2022

Publisher's Note: MDPI stays neutral with regard to jurisdictional claims in published maps and institutional affiliations.

Abstract: (1) Background: Hypofractionated stereotactic radiotherapy (HSRT) and anti-vascular endothelial growth factor (VEGF) antibodies have been reported to have a promising survival benefit in recent studies. Anlotinib is a new oral VEGF receptor inhibitor. This report describes our experience using HSRT and anlotinib for recurrent glioblastoma (rGBM). (2) Methods: Between December 2019 and June 2020, rGBM patients were retrospectively analysed. Anlotinib was prescribed at 12 mg daily during HSRT. Adjuvant anlotinib was administered d1-14 every 3 weeks. The primary endpoint was the objective response rate (ORR). Secondary endpoints included overall survival (OS), progression-free survival (PFS) after salvage treatment, and toxicity. (3) Results: Five patients were enrolled. The prescribed dose was 25.0 Gy in 5 fractions. The median number of cycles of anlotinib was 21 (14–33). The ORR was 100%. Three (60%) patients had the best outcome of a partial response (PR), and 2 (40%) achieved a complete response (CR). One patient died of tumour progression at the last follow-up. Two patients had grade 2 hand-foot syndrome. (4) Conclusions: Salvage HSRT combined with anlotinib showed a favourable outcome and acceptable toxicity for rGBM. A prospective phase II study (NCT04197492) is ongoing to further investigate the regimen.

Keywords: hypofractionated stereotactic radiotherapy; recurrent high-grade glioma; salvage treatment; anlotinib

1. Introduction

Glioblastoma is the most frequently diagnosed malignant primary brain tumour in adults. Maximum surgical resection with six courses of temozolomide adjuvant chemoradiotherapy is the current standard of care in the first-line management of glioblastoma [1]. However, most patients still suffer from recurrence within eight months after primary treatment, and approximately 90% of recurrences occur within a 2 cm margin of the original tumour resection cavity [2]. The management of recurrent glioblastoma is highly challenging due to resistance to available therapeutic approaches, and treatment outcomes remain uniformly poor.

Table 1. Patient characteristics and treatment outcomes.

Case	Age Sex	Interval between Initial Diagnosis and HSRT (Months)	Upfront RT Dose/fx	Upfront Chemotherapy (Cycles)	MGMT	IDH1	1p/19q
1	60 Male	12.6	60Gy/30	TMZ (12)	+	-	-
2	46 Female	10.4	60Gy/30	TMZ (6)	+	-	-
3	55 Female	14.8	60Gy/30	TMZ (12)	+	-	+
4	51 Male	10.0	60Gy/30	TMZ (4)	+	-	-
5	43 Male	7.0	60Gy/30	TMZ (4)	+	-	-

Case	TERT	Recurrent Lesion	Recurrent PTV (cm^3)	KPS at HSRS	Dose (iso-dose line)	Cycles of Anlotinib	F/U Interval from HSRS (months)
1	+	Left Frontal Lobe	7.08	80	68	15	10
2	+	Left Frontal Lobe	26.94	80	65	14	10
3	+	Left Occipital Lobe	54.41	70	70	9	6
4	+	Right Frontal Lobe	5.53	90	70	4	4
5	+	Left and Right Frontal Lobe	44.33	90	68	8	6

Table 2. Best treatment outcomes and adverse events that occurred in rGBM patients.

Outcomes/AE	Total No. of Patients	No. of Patients		
		Grade 1	Grade 2	Grade 3
ORR	5 (100%)			
CR	2		N/A	
PR	3			
Haematologic				
Thrombocytopenia	1	0	1	0
Nonhaematologic				
Hand foot syndrome	2	0	2	0
Rash	1	0	1	0
Hypertension	1	0	1	0

3.3. Treatment Outcomes

The patients were assessed by the RANO criteria. Three (60%) patients had the best outcome of PR, and two (40%) achieved CR; the ORR was 100% (Figure 1A). The follow-up from the time of HSRT ranged from 20 to 26 months. By the end of the study, four patients had progressive disease (PD) and one patient died of tumour progression (Figure 1B). The overall survival rates following the salvage treatment were 100% and 80% at one and two years, and the PFS rates were 60% and 40%, respectively (see Supplementary Figures S1 and S2).

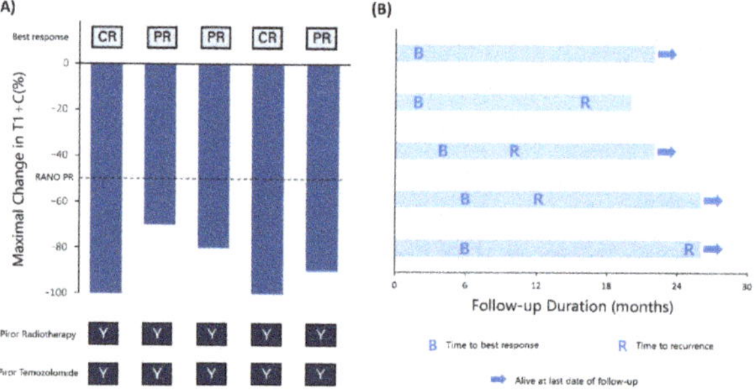

Figure 1. (**A**) Maximal change in the product of the perpendicular diameter in MRI T1 contrast before and after HSRT with anlotinib in each patient. CR, complete response. PR, partial response. RANO, Response Assessment in Neuro-Oncology. Y, yes. (**B**) Follow-up duration, time to the best response, and time to recurrence in each patient.

4. Discussion

Recurrent glioblastoma has been reported to have a poor prognosis. Due to its therapeutic resistance and aggressiveness, its clinical management is challenging. GBM is a vascularised tumour that produces VEGF. Anti-VEGF treatments have been widely used in recurrent GBM. The mechanism of anti-VEGF treatments may have two aspects. First, inhibiting VEGF and its receptor reduces tumour angiogenesis to produce a hypoxic environment and inhibits tumour growth [6]. Second, the tumour vessel diameter was normalised, and the basement membrane was thin. A reduced volume of tumour microvessels has been reported to be related to longer survival [7].

Bevacizumab is approved for treating recurrent glioblastoma by the US Food and Drug Administration and has become a recommended treatment in the National Comprehensive Cancer Network (NCCN) guidelines, with several phase II and III randomised trials indicating a prolonged PFS compared with chemotherapy alone [8–10]. A phase III RCT reported a prolonged median PFS (4.2 vs. 1.5 months) in the bevacizumab and Lomustine groups compared with the Lomustine alone group. However, this trial did not find a difference in OS between the two groups. The grade 3 to 5 toxicity rate in the experimental group was 63.6% [10]. Friedman et al. reported a phase II randomised controlled trial (RCT) in which a higher 6-month PFS rate (50.3% vs. 42.6%) and better ORR (37.8% vs. 28.2%) were observed in the bevacizumab with irinotecan group than in the bevacizumab alone group [9]. Other anti-angiogenic drugs, including sorafenib, pazopanib, sunitinib, etc., were reported in phase I and II trials treating rGBM. The ORR reported for anti-VEGF treatments for rGBM ranged from 6% to 30% (Table 3), and the 6-month PFS ranged from 3% to 63%. The treatment-related toxicity was mild for these anti-VEGF treatments. However, the efficacy seems to be unsatisfactory.

Anlotinib is an oral novel multi-target tyrosine kinase inhibitor targeting the VEGF1/2/3 receptor, fibroblast growth factor receptor and platelet-derived growth factor receptor. It inhibits more targets than bevacizumab, sunitinib, sorafenib, etc. and has been reported to reduce both tumour proliferation and angiogenesis [3]. Lv et al. published the first case report of the administration of 12 mg anlotinib to an rGBM patient. The patient achieved a PR after 26 days, but the tumour progressed in two months [4]. Wang et al. reported a recurrent GBM patient with an FGFR-TACC3 fusion who was administered anlotinib 12 mg and temozolomide 100 mg/m^2. The patient achieved a PR after two months and maintained stable disease for more than 17 months [11].

Several reports have suggested that re-irradiation has a reasonable efficacy with acceptable safety profiles in selected patients with recurrent GBM. However, for rGBM, salvage treatment failure ultimately occurs. It is crucial to increase local treatment to reduce recurrence risk. In a meta-analysis, a highly conformal technique with a hypofractionated regimen (e.g., 25 Gy in five fractions or 35 Gy in 10 fractions) is recommended, considering the volume and location of the recurrent tumour. The RTOG 1205 trial reported a prolonged PFS with anti-VEGF treatment with HSRT compared with bevacizumab alone [5]. Philip et al. theorized that additional anti-VEGF treatment sensitised the tumour endothelia to radiotherapy and induced apoptosis [12]. New-generation automated noncoplanar HSRT delivery systems can deliver high-dose treatment by limiting the dose to normal structures and can provide a higher local treatment intensity for recurrent tumours. In this study, the ORR rate of salvage treatment was 100% in two CR and three PR patients. The ORR was higher than other results of anti-VEGF treatments [13–25], which ranged from 6% to 30% (Table 3). There may be several possible reasons for the promising treatment outcomes. Patient selection may be a reason for good outcomes. All patients had a KPS of 70 or higher, and HSRT was performed after the first recurrence. Moreover, the administration of HSRT increased the local treatment intensity. The preliminary result of RTOG 1205 also reported an increased PFS in the intensified treatment groups. Additionally, patients with a smaller tumour volume may have a better response. The two CR patients (Figure 2A,B) in this study had a relatively smaller PTV (7.08 and 5.53 cm^3) than the three PR patients (26.94, 44.33, and 54.41 cm^3).

Table 3. Reported anti-angiogenic treatment for recurrent glioblastoma.

Author, Year	Treatment	Phase (Sample Size)	Outcome (ORR Rate%)	Median PFS (Months)	Median OS (Months)	6-Month PFS
Reardon, 2018 [13]	Trebananib	II (11)	2CR (18)	0.7	11.4	N/A
Reardon, 2005 [14]	Imatinib	II (33)	3PR (9)	3.3	N/A	27.0%
Iwamoto, 2010 [15]	Pazopanib	II (35)	8PR (22)	3.0	8.1	3.0%
Pan, 2012 [16]	Sunitinib	II (16)	0	N/A	12.6	16.7%
Hutterer, 2014 [17]	Sunitinib	II (40)	0	2.0	9.2	12.5%
Hassler, 2014 [18]	Imatinib	II (24)	2PR (8)	3.0	6.2	N/A
Batchelor, 2010 [19]	Cediranib	II (131)	1CR, 17PR (14)	3.0	8.0	16.0%
Gerstner, 2015 [20]	Cediranib	I (45)	2CR, 2PR (9)	1.9	6.5	4.4%
Chheda, 2015 [21]	Vandetanib	I (19)	2PR (11)	1.9	7.2	63%
McNeill, 2014 [22]	Vandetanib	II (32)	2PR (6)	1.7	5.6	N/A
Duerinck, 2016 [23]	Axitinib	II (22)	2CR, 4PR (27)	N/A	6.7	34%
Lee, 2012 [24]	Sorafenib	I/II (18)	2PR (11)	1.8	N/A	N/A
Groot, 2020 [25]	Aflibercept	II (27)	8PR (30)	N/A	N/A	N/A

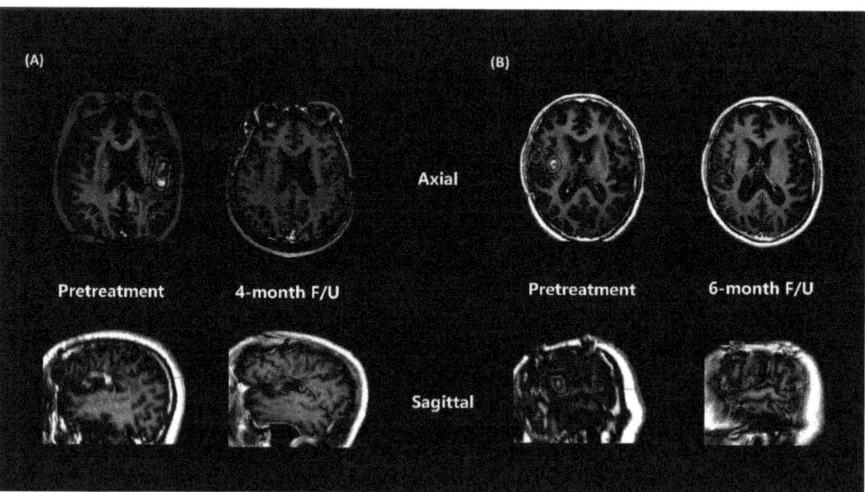

Figure 2. Contrast-enhanced MRI T1 of responses to HSRT and anlotinib, including (**A**) patient case 1 and (**B**) patient case 4, who achieved a complete response.

Salvage HSRT was administered with a full dose of 25 Gy/5 fx for all five patients without any interruption. No radiation necrosis occurred during the follow-up. Grade 2 hand-food syndrome was found in two patients (40%), and rash and hypertension were observed in one patient (20%). These adverse effects were considered to be related to anlotinib. In a phase II randomised trial of non-small-cell lung cancer patients, 28.33% of the subjects had grade 2 hand-foot syndrome, and grade 2 hypertension was observed in 55% of patients [26]. These toxicities were also observed in our study.

The study had some limitations due to its retrospective nature: an inherent patient selection bias was created when the physicians chose eligible patients to receive the regimen. The treatment option was provided for patients with high KPS scores who were not willing to receive standard intravenous bevacizumab treatment. Thus, the cohort was enriched with patients with a better prognosis. Another limitation was that recurrence before salvage treatment was diagnosed by radiological parameters according to the RANO criteria, which is a common practice [27]. However, the lack of biopsy samples limited the information on tumour genomic characterisations. It is crucial to consider whether the previously detected mutation still presents as the dominant clone at the time of recurrence [28]. Further investigation is warranted to explain the potential treatment mechanisms and select good responders to the regimen.

Despite the limitations, this study provides initial evidence of a promising outcome using salvage HSRT with anlotinib in a real-world scenario. Responses were observed in all rGBM patients included in the study. Further investigation is needed to identify patients who can benefit from this regimen. A prospective phase II study HSCK-002 (ClinicalTrials.gov identifier: NCT04197492) is ongoing to further investigate the value of HSRT with anlotinib.

5. Conclusions

Salvage radiosurgery with anlotinib appeared to achieve a clinical benefit with acceptable toxicity for rGBM patients in this preliminary report. A prospective phase II study (NCT04197492) is ongoing to further investigate the value of HSRT with anlotinib in rHGG.

Supplementary Materials: The following supporting information can be downloaded at: https://www.mdpi.com/article/10.3390/brainsci12040471/s1. Figure S1: Overall survival (OS) from salvage treatment of all rHGG patients (calculated with the Kaplan–Meier method). Figure S2: Progression-free survival (PFS) from salvage treatment of all rHGG patients (calculated with the Kaplan–Meier method).

Author Contributions: Conceptualization, E.W., X.W., Y.L. and Y.G.; methodology, E.W., X.W. and Y.G.; software, Y.G.; validation, E.W., X.W. and Y.L.; formal analysis, Y.G.; investigation, L.P., J.D. and Y.W.; resources, L.P., E.W. and X.W.; data curation, J.L., X.G., H.Z., C.L., G.M. and X.L.; writing—original draft preparation, Y.G., J.L., X.G.; writing—review and editing, X.W., Y.G. and X.G.; visualization, Y.G.; supervision, E.W.; project administration, X.W.; funding acquisition, X.W. All authors have read and agreed to the published version of the manuscript.

Funding: This research was funded by the National Natural Science Foundation of China grant number 81727806.

Institutional Review Board Statement: The study was conducted in accordance with the Declaration of Helsinki and was approved by the ethics committee of Huashan Hospital.

Informed Consent Statement: Informed consent was obtained from all subjects involved in the study.

Data Availability Statement: Research data are stored in an institutional repository and will be shared upon request to the corresponding author.

Conflicts of Interest: The authors declare no conflict of interest.

References

1. Attarian, F.; Taghizadeh-Hesary, F.; Fanipakdel, A.; Javadinia, S.A.; Porouhan, P.; PeyroShabany, B.; Fazilat-Panah, D. A Systematic Review and Meta-Analysis on the Number of Adjuvant Temozolomide Cycles in Newly Diagnosed Glioblastoma. *Front. Oncol.* **2021**, *11*, 779491. [CrossRef] [PubMed]
2. Sneed, P.K.; Gutin, P.H.; Larson, D.A.; Malec, M.K.; Phillips, T.L.; Prados, M.D.; Scharfen, C.O.; Weaver, K.A.; Wara, W.M. Patterns of Recurrence of Glioblastoma Multiforme after External Irradiation Followed by Implant Boost. *Int. J. Radiat. Oncol. Biol. Phys.* **1994**, *29*, 719–727. [CrossRef]
3. Shen, G.; Zheng, F.; Ren, D.; Du, F.; Dong, Q.; Wang, Z.; Zhao, F.; Ahmad, R.; Zhao, J. Anlotinib: A Novel Multi-Targeting Tyrosine Kinase Inhibitor in Clinical Development. *J. Hematol. Oncol.* **2018**, *11*, 120. [CrossRef] [PubMed]
4. Lv, Y.; Zhang, J.; Liu, F.; Song, M.; Hou, Y.; Liang, N. Targeted Therapy with Anlotinib for Patient with Recurrent Glioblastoma: A Case Report and Literature Review. *Medicine* **2019**, *98*, e15749. [CrossRef]
5. Tsien, C.; Pugh, S.; Dicker, A.P.; Raizer, J.J.; Matuszak, M.M.; Lallana, E.; Huang, J.; Algan, O.; Taylor, N.; Portelance, L. Randomized Phase II Trial of Re-Irradiation and Concurrent Bevacizumab versus Bevacizumab Alone as Treatment for Recurrent Glioblastoma (NRG Oncology/RTOG 1205): Initial Outcomes and RT Plan Quality Report. *Int. J. Radiat. Oncol. Biol. Phys.* **2019**, *105*, S78. [CrossRef]
6. Wong, E.T.; Gautam, S.; Malchow, C.; Lun, M.; Pan, E.; Brem, S. Bevacizumab for Recurrent Glioblastoma Multiforme: A Meta-Analysis. *J. Natl. Compr. Cancer Netw.* **2011**, *9*, 403–407. [CrossRef]
7. Sorensen, A.G.; Batchelor, T.T.; Zhang, W.-T.; Chen, P.-J.; Yeo, P.; Wang, M.; Jennings, D.; Wen, P.Y.; Lahdenranta, J.; Ancukiewicz, M. A "Vascular Normalization Index" as Potential Mechanistic Biomarker to Predict Survival after a Single Dose of Cediranib in Recurrent Glioblastoma Patients. *Cancer Res.* **2009**, *69*, 5296–5300. [CrossRef]

8. Vredenburgh, J.J.; Desjardins, A.; Herndon, J.E.; Marcello, J.; Reardon, D.A.; Quinn, J.A.; Rich, J.N.; Sathornsumetee, S.; Gururangan, S.; Sampson, J. Bevacizumab plus Irinotecan in Recurrent Glioblastoma Multiforme. *J. Clin. Oncol.* **2007**, *25*, 4722–4729. [CrossRef]
9. Friedman, H.S.; Prados, M.D.; Wen, P.Y.; Mikkelsen, T.; Schiff, D.; Abrey, L.E.; Yung, W.A.; Paleologos, N.; Nicholas, M.K.; Jensen, R. Bevacizumab Alone and in Combination with Irinotecan in Recurrent Glioblastoma. *J. Clin. Oncol.* **2009**, *27*, 4733–4740. [CrossRef]
10. Wick, W.; Gorlia, T.; Bendszus, M.; Taphoorn, M.; Sahm, F.; Harting, I.; Brandes, A.A.; Taal, W.; Domont, J.; Idbaih, A. Lomustine and Bevacizumab in Progressive Glioblastoma. *N. Engl. J. Med.* **2017**, *377*, 1954–1963. [CrossRef]
11. Wang, Y.; Liang, D.; Chen, J.; Chen, H.; Fan, R.; Gao, Y.; Gao, Y.; Tao, R.; Zhang, H. Targeted Therapy with Anlotinib for a Patient with an Oncogenic FGFR3-TACC3 Fusion and Recurrent Glioblastoma. *Oncologist* **2021**, *26*, 173–177. [CrossRef] [PubMed]
12. Gutin, P.H.; Iwamoto, F.M.; Beal, K.; Mohile, N.A.; Karimi, S.; Hou, B.L.; Lymberis, S.; Yamada, Y.; Chang, J.; Abrey, L.E. Safety and Efficacy of Bevacizumab with Hypofractionated Stereotactic Irradiation for Recurrent Malignant Gliomas. *Int. J. Radiat. Oncol. Biol. Phys.* **2009**, *75*, 156–163. [CrossRef] [PubMed]
13. Reardon, D.A.; Lassman, A.B.; Schiff, D.; Yunus, S.A.; Gerstner, E.R.; Cloughesy, T.F.; Lee, E.Q.; Gaffey, S.C.; Barrs, J.; Bruno, J. Phase 2 and Biomarker Study of Trebananib, an Angiopoietin-blocking Peptibody, with and without Bevacizumab for Patients with Recurrent Glioblastoma. *Cancer* **2018**, *124*, 1438–1448. [CrossRef] [PubMed]
14. Reardon, D.A.; Egorin, M.J.; Quinn, J.A.; Rich, J.N.; Gururangan, I.; Vredenburgh, J.J.; Desjardins, A.; Sathornsumetee, S.; Provenzale, J.M.; Herndon, J.E. Phase II Study of Imatinib Mesylate plus Hydroxyurea in Adults with Recurrent Glioblastoma Multiforme. *J. Clin. Oncol.* **2005**, *23*, 9359–9368. [CrossRef]
15. Iwamoto, F.M.; Lamborn, K.R.; Robins, H.I.; Mehta, M.P.; Chang, S.M.; Butowski, N.A.; DeAngelis, L.M.; Abrey, L.E.; Zhang, W.-T.; Prados, M.D. Phase II Trial of Pazopanib (GW786034), an Oral Multi-Targeted Angiogenesis Inhibitor, for Adults with Recurrent Glioblastoma (North American Brain Tumor Consortium Study 06-02). *Neuro-Oncology* **2010**, *12*, 855–861. [CrossRef]
16. Pan, E.; Yu, D.; Yue, B.; Potthast, L.; Chowdhary, S.; Smith, P.; Chamberlain, M. A Prospective Phase II Single-Institution Trial of Sunitinib for Recurrent Malignant Glioma. *J. Neuro-Oncol.* **2012**, *110*, 111–118. [CrossRef]
17. Hutterer, M.; Nowosielski, M.; Haybaeck, J.; Embacher, S.; Stockhammer, F.; Gotwald, T.; Holzner, B.; Capper, D.; Preusser, M.; Marosi, C. A Single-Arm Phase II Austrian/German Multicenter Trial on Continuous Daily Sunitinib in Primary Glioblastoma at First Recurrence (SURGE 01-07). *Neuro-Oncology* **2014**, *16*, 92–102. [CrossRef]
18. Hassler, M.R.; Vedadinejad, M.; Flechl, B.; Haberler, C.; Preusser, M.; Hainfellner, J.A.; Wöhrer, A.; Dieckmann, K.U.; Rössler, K.; Kast, R. Response to Imatinib as a Function of Target Kinase Expression in Recurrent Glioblastoma. *SpringerPlus* **2014**, *3*, 111. [CrossRef]
19. Batchelor, T.T.; Duda, D.G.; di Tomaso, E.; Ancukiewicz, M.; Plotkin, S.R.; Gerstner, E.; Eichler, A.F.; Drappatz, J.; Hochberg, F.H.; Benner, T. Phase II Study of Cediranib, an Oral Pan–Vascular Endothelial Growth Factor Receptor Tyrosine Kinase Inhibitor, in Patients with Recurrent Glioblastoma. *J. Clin. Oncol.* **2010**, *28*, 2817. [CrossRef]
20. Gerstner, E.R.; Ye, X.; Duda, D.G.; Levine, M.A.; Mikkelsen, T.; Kaley, T.J.; Olson, J.J.; Nabors, B.L.; Ahluwalia, M.S.; Wen, P.Y. A Phase I Study of Cediranib in Combination with Cilengitide in Patients with Recurrent Glioblastoma. *Neuro-Oncology* **2015**, *17*, 1386–1392. [CrossRef]
21. Chheda, M.G.; Wen, P.Y.; Hochberg, F.H.; Chi, A.S.; Drappatz, J.; Eichler, A.F.; Yang, D.; Beroukhim, R.; Norden, A.D.; Gerstner, E.R. Vandetanib plus Sirolimus in Adults with Recurrent Glioblastoma: Results of a Phase I and Dose Expansion Cohort Study. *J. Neuro-Oncol.* **2015**, *121*, 627–634. [CrossRef] [PubMed]
22. McNeill, K.; Iwamoto, F.; Kreisl, T.; Sul, J.; Shih, J.; Fine, H. AT-39A Randomized Phase II Trial of Vandetanib (ZD6474) in Combination with Carboplatin versus Carboplatin Alone in Adults with Recurrent Glioblastoma. *Neuro-Oncology* **2014**, *16* (Suppl. S5), v17. [CrossRef]
23. Duerinck, J.; Du Four, S.; Vandervorst, F.; D'Haene, N.; Le Mercier, M.; Michotte, A.; Van Binst, A.M.; Everaert, H.; Salmon, I.; Bouttens, F. Randomized Phase II Study of Axitinib versus Physicians Best Alternative Choice of Therapy in Patients with Recurrent Glioblastoma. *J. Neuro-Oncol.* **2016**, *128*, 147–155. [CrossRef]
24. Lee, E.Q.; Kuhn, J.; Lamborn, K.R.; Abrey, L.; DeAngelis, L.M.; Lieberman, F.; Robins, H.I.; Chang, S.M.; Yung, W.A.; Drappatz, J. Phase I/II Study of Sorafenib in Combination with Temsirolimus for Recurrent Glioblastoma or Gliosarcoma: North American Brain Tumor Consortium Study 05-02. *Neuro-Oncology* **2012**, *14*, 1511–1518. [CrossRef] [PubMed]
25. De Groot, J.F.; Wen, P.Y.; Lamborn, K.; Chang, S.; Cloughesy, T.F.; Chen, A.P.; DeAngelis, L.M.; Mehta, M.P.; Gilbert, M.R.; Yung, W.K. Phase II Single Arm Trial of Aflibercept in Patients with Recurrent Temozolomide-Resistant Glioblastoma: NABTC 0601. *J. Clin. Oncol.* **2008**, *26* (Suppl. S15), 2020. [CrossRef]
26. Han, B.; Li, K.; Zhao, Y.; Li, B.; Cheng, Y.; Zhou, J.; Lu, Y.; Shi, Y.; Wang, Z.; Jiang, L. Anlotinib as a Third-Line Therapy in Patients with Refractory Advanced Non-Small-Cell Lung Cancer: A Multicentre, Randomised Phase II Trial (ALTER0302). *Br. J. Cancer* **2018**, *118*, 654–661. [CrossRef]
27. Vogelbaum, M.A.; Jost, S.; Aghi, M.K.; Heimberger, A.B.; Sampson, J.H.; Wen, P.Y.; Macdonald, D.R.; Van den Bent, M.J.; Chang, S.M. Application of Novel Response/Progression Measures for Surgically Delivered Therapies for Gliomas: Response Assessment in Neuro-Oncology (RANO) Working Group. *Neurosurgery* **2012**, *70*, 234–244. [CrossRef]
28. Kaley, T.; Touat, M.; Subbiah, V.; Hollebecque, A.; Rodon, J.; Lockhart, A.C.; Keedy, V.; Bielle, F.; Hofheinz, R.-D.; Joly, F. BRAF Inhibition in BRAFV600-Mutant Gliomas: Results from the VE-BASKET Study. *J. Clin. Oncol.* **2018**, *36*, 3477. [CrossRef]

brain sciences

MDPI

Article

Comparison of Immune Checkpoint Molecules *PD-1* and *PD-L1* in Paired Primary and Recurrent Glioma: Increasing Trend When Recurrence

Wei Yu [1],[†], Anwen Shao [2],[†], Xiaoqiu Ren [1], Zexin Chen [3], Jinghong Xu [4] and Qichun Wei [1],[*]

[1] Department of Radiation Oncology, Key Laboratory of Cancer Prevention and Intervention, Ministry of Education, Second Affiliated Hospital, Zhejiang University School of Medicine, Hangzhou 310009, China; wei-yu@zju.edu.cn (W.Y.); rxq@zju.edu.cn (X.R.)

[2] Department of Neurosurgery, The Second Affiliated Hospital, Zhejiang University School of Medicine, Hangzhou 310009, China; 2316040@zju.edu.cn

[3] Center of Clinical Epidemiology and Biostatistics for Statistical Analysis, The Second Affiliated Hospital, Zhejiang University School of Medicine, Hangzhou 310009, China; chenzexin@zju.edu.cn

[4] Department of Pathology, The Second Affiliated Hospital, Zhejiang University School of Medicine, Hangzhou 310009, China; zydxjh@zju.edu.cn

[*] Correspondence: qichun_wei@zju.edu.cn; Tel.: +86-571-8778-3522; Fax: +86-571-8721-4404

[†] These authors contributed equally to this article.

Abstract: Purpose: This study aims to investigate *PD-1/PD-L1* expression patterns in paired primary and recurrent gliomas. **Methods:** From January 2008 to December 2014, 42 patients who underwent surgical resections of primary and recurrent gliomas were retrospectively included. *PD-1/PD-L1* protein expression in tumors was evaluated through immunohistochemistry. **Results:** In primary gliomas, *PD-1* and *PD-L1* expression was evident in 9 (22.0%) and 14 (33.3%) patients. In the paired recurrent glioma, *PD-1* and *PD-L1* expression was evident in 25 (61.0%) and 31 (74.0%) lesions. Both *PD-1* and *PD-L1* showed significantly enhanced expression after recurrence ($p < 0.005$; $p < 0.005$). For *PD-L1* expression in recurrent gliomas, the adjuvant therapy group showed significantly increased expression compared to primary gliomas ($p < 0.005$). For *PD-1*- primary gliomas, if the matched recurrent gliomas showed *PD-1+*, the PFS became worse than the remaining recurrent gliomas *PD-1*- (12.7 vs. 25.9 months, $p = 0.032$). Interestingly, for *PD-L1*- primary gliomas, if the matched recurrent gliomas showed *PD-L1+*, the OS became better than the remaining recurrent gliomas *PD-L1*- (33.8 vs. 17.5 months, $p < 0.001$). **Conclusions:** In the study, we found the expression of *PD-1/PD-L1* increased significantly in recurrent gliomas and the elevated level of *PD-L1* was tightly associated with adjuvant treatment, suggesting the potential therapeutic and predictive value of *PD-1* and *PD-L1* in the treatment of recurrent gliomas.

Keywords: glioma; recurrence; programmed cell death 1 receptor; programmed death-ligand 1; therapeutics

Citation: Yu, W.; Shao, A.; Ren, X.; Chen, Z.; Xu, J.; Wei, Q. Comparison of Immune Checkpoint Molecules *PD-1* and *PD-L1* in Paired Primary and Recurrent Glioma: Increasing Trend When Recurrence. *Brain Sci.* **2022**, *12*, 266. https://doi.org/10.3390/brainsci12020266

Academic Editors: Hailiang Tang and Sergey Kasparov

Received: 22 December 2021
Accepted: 3 February 2022
Published: 14 February 2022

Publisher's Note: MDPI stays neutral with regard to jurisdictional claims in published maps and institutional affiliations.

1. Introduction

Glioma accounts for 36.7–42.6% of all primary brain tumors, which is the most common primary brain neoplasm [1–3]. Gliomas can be histopathologically classified as a low-grade glioma or high-grade glioma. The median survival time for a low-grade glioma is 5–10 years. However, 50–75% of low-grade gliomas will eventually develop into a high-grade glioma. Glioblastoma is highly malignant and shows rapid clinical progression. Notably, glioblastoma accounts for the highest proportion of all gliomas [1,2]. The aggressive treatments on glioblastoma include surgical resection with a maximum safety margin, followed by concurrent chemoradiotherapy and six cycles of adjuvant temozolomide chemotherapy. Even after receiving all these treatments, the median survival time of patients with newly diagnosed glioblastoma is around 14.6 months [4–6]. Despite effective radiotherapy and chemotherapy after the operation, most patients will inevitably relapse after treatment. The recurrent glioma is more aggressive, progresses more rapidly, and resists treatment [7]. The treatment of recurrent gliomas has always been a challenge for clinicians. These years, the anti-tumor effects of immune checkpoint inhibitors have been increasingly recognized. They are highly reactive in bladder carcinoma, head and neck cancer, melanoma, lung cancer, and renal cell carcinoma with long-term tumor remission [8–11]. The brain is an organ with a unique tumor immune microenvironment [12]. The presence of tumor-infiltrating lymphocytes (TILs) in tumor lesions indicates that the immune modulation is involved in brain neoplasm progression [13]. Tumors generate immunosuppressive systems to avoid host immune attacks [14]. *Programmed death protein 1 (PD-1)* and *programmed death-ligand 1 (PD-L1)* play critical roles in tumor immune escape [15]. In 2000, Freeman et al. confirmed that *PD-1* could reverse modulate lymphocyte activation [16]. *PD-1*, a co-inhibitory receptor, is expressed in activated *CD4+* or *CD8+* T cells. *PD-L1*, the ligand of *PD-1*, is represented by a variety of cells, including tumor cells and lymphocytes. The reaction between *PD-1* and *PD-L1* in T cells could exhaust T cell clones, allowing tumor cells to survive under normal physiological conditions [17–19].

PD-L1 expression in tumors, validated by immunohistochemistry (IHC) assays, is related to the response of *PD-L1/PD-1* inhibitors. Besides, it is suggested as a biomarker in clinical practice [9,20]. Multiple studies show the evidence that *PD-1/PD-L1* is expressed in gliomas [21–23]. However, up to now, few systematic studies have compared the expression of immune checkpoint molecule *PD-1/PD-L1* genes in primary and matched recurrent specimens of glioma.

Data comparing paired primary and recurrent gliomas are scarce, due to the small proportion of reoperation in recurrent gliomas. In this study, 42 patients who underwent surgical resections of both primary and recurrent gliomas were retrospectively included. *PD-1/PD-L1* expression in tumors was assessed by IHC. This comparative study will help us better understand the tumor biology of glioma, thus providing new insights into the treatment strategy.

2. Materials and Methods

2.1. Patients

From January 2008 to December 2015, 42 patients who received surgical resections of both primary and recurrent gliomas in the Second Affiliated Hospital of Zhejiang University (SAHZU) were retrospectively included. Histopathology diagnosis was acquired after operations and diagnosed by neuropathologists. Patients were included if they met the following inclusion criteria: (1) patients received operations of both primary and recurrent gliomas; (2) primary and recurrent gliomas were histopathologically confirmed; (3) long-term follow-up in SAHZU after treatment. On the other hand, patients were excluded based on the following: (1) patients received immune therapy; (2) one of the operations was not conducted in SAHZU; (3) lack of pathological confirmation. *PD-1*, *PD-L1* and *IDH1* expression in tumors was assessed by IHC. This study was approved by the Ethics Committee of SAHZU, and was carried out in accordance with the Declaration of Helsinki. The code of ethics was 046.

2.2. Immunohistochemistry

The primary antibodies used in this study were *PD-1* (Abcam, Cambridge, UK), *PD-L1* (Cell Signaling Technology, Danvers, MA, USA) and *IDH1* (ZSGB-BIO, Beijing, China). Formalin-Fixed and Paraffin-Embedded (FFPE) blocks from the pathology department of SAHZU were cut into serial 4 um slices with a microtome. The paraffin sections were incubated at 60 °C in the incubator overnight. The sections were then deparaffinized in xylene and rehydrated through graded alcohols (100%, 95%, 75%). The sections for *PD-1*, *PD-L1*, *IDH1* testing were put in boiled SignalStain EDTA Unmasking Solution (pH 9) in the electric cooker for 10 min and heat-preserved for 10 min. Then the sections were blocked with 1% BSA (Bovine serum Albumin) and incubated at 4 °C overnight with primary antibody (the dilution rate of the primary antibody was as follows: *PD-1* (1:50), *PD-L1* (1:200), *IDH1* (ready to use). The day after, sections were incubated with HRP-conjugated secondary antibody at room temperature (the second antibody used for *PD-1* and *IDH1* was the Polink-1 HRP staining system (ZSGB-BIO), incubated for 15 min. The second antibody used for *PD-L1* was SignalStain Boost IHC Detection Reagent (Cell Signaling Technology), incubated for 30 min). After that, the sections were incubated with chromagen 3, 3′-diaminobenzidine (DAB) at room temperature (10 min for *PD-1*, 2 min for *PD-L1* and *IDH1*). The sections were counterstained with hematoxylin for 3 min and dehydrated through graded alcohols (75%, 95%, 100%) and mounted.

2.3. Scoring System

PD-1 was expressed in lymphocytes, so the immunohistochemical staining evaluation of the *PD-1* protein was unique: to find the area with the highest density of TILs in low magnification, counting *PD-1* expression cell at a $400\times$ visual field (/HPF). Staining was scored as positive if the number of *PD-1*-positive TIL was one cell per high-power field [24,25]. Evaluation of *PD-L1* protein expression by IHC staining was based on the range and the intensity. Briefly, the proportion of immunopositive cells among the total number of tumor cells was subdivided into five categories, as follows: 0, <1%; 1, 1–25%; 2, 25–50%; 3, 50–75%; and 4, >75% positive cells. The immunointensity was subclassified into four groups: 0, negative; 1, weak; 2, moderate; and 3, strong immunointensity. IHC scores were generated by multiplication of these two parameters. The value of multiplication less than two was defined as *PD-L1*-negative. The value of multiplication equal to or more than two was defined as *PD-L1*-positive. Immunohistochemical staining evaluation of the IDH1 protein was based on the intensity of staining. Negative staining was defined as wild-type, and positive staining was defined as a mutation.

2.4. Statistical Analysis

A normality test (Kolmogorov–Smirnov test) and homogeneity of variance test (Levene test) were conducted in measurement data, such as age, among subgroups. In each test, $p > 0.1$ was considered for the Gaussian distribution and homogeneity of variance, respectively. A two-group t-test was conducted among subgroups to compare the measurement data if they had a Gaussian distribution and homogeneity of variance. The chi-square test and Fisher's exact test was conducted among subgroups to compare enumeration data, such as sex, tumor location, WHO grade, residue, adjuvant therapy, and type of recurrence. $p < 0.05$ was considered significant. For the ranked data of paired primary and recurrent samples, the Wilcoxon rank-sum test was used to analyze the variation trend of each gene expression. Correlations between *PD-1* and *PD-L1* gene expression were tested by the Spearman rank correlation test. Overall survival (OS) is defined as the time from the first surgical treatment until the death of the patient. Progression-free survival (PFS) is defined as the time from the first operation of the glioma to the progression of the tumor. PFS and OS were estimated by the Kaplan–Meier method with a two-sided log-rank test. $p < 0.05$ was considered statistically significant. IBM SPSS Statistics 26.0 was used for statistical analyses.

3. Results

3.1. Patients' Characteristics and Follow-Up

From January 2008 to December 2015, 42 patients who received surgical resections of both primary and recurrent gliomas were retrospectively included (Figure 1). The genders of the patients were 27 male and 15 female. The average age at first operation was 43.2 years (range: 11–61 years old). After the first surgical resection, 21 cases had residual tumors, and 21 had no residual tumor. There were two pilocytic astrocytomas, eight astrocytomas with IDH1-mutant, 15 astrocytomas with IDH1-wildtype, three oligodendrogliomas, NOS, 12 glioblastomas with IDH1-wildtype, and one ganglioglioma. Nine patients were IDH1-mutant, 33 were IDH1-wildtype. Six patients received radiation alone, one patient received only chemotherapy, 15 patients received radiation and chemotherapy, and 20 patients had not received adjuvant therapy. Before the secondary operation, 38 cases recurred in situ, and four cases had a distant recurrence in the brain. After the second operation, 31 patients had residual tumors, and 11 patients had no residual tumor. There was one pilocytic astrocytoma, nine astrocytomas with IDH1-mutant, 10 astrocytomas with IDH1-wildtype, four oligodendrogliomas, NOS, and 18 glioblastomas with IDH1-wildtype. Eleven patients were IDH1-mutant, 31 were IDH1-wildtype. After that, one patient received only radiation, eight patients received chemotherapy alone, nine patients received radiotherapy and chemotherapy, and 24 patients had not received adjuvant treatment. The last follow-up time was 18 August 2017, with a median PFS of 13.5 months (95% CI: 11.0–16.05 months). The median PFS of newly diagnosed low-grade gliomas was 18 months (95% CI: 12.4–24.6 months), and the median PFS of the newly diagnosed high-grade gliomas was 12.7 months (95% CI: 10.0–15.4 months). At the time of the last follow-up, 12 patients were alive. The median OS was 33.8 months (95% CI: 28.3–39.3) for all 42 patients. The median OS of newly diagnosed low-grade gliomas was 54.9 months (95% CI: 30.9–78.9), and the median OS of the newly diagnosed high-grade gliomas was 30.0 months (95% CI: 22.2–37.8): 31.0 months (95% CI: 26.6–35.4) for grade III gliomas and 24.4 months (95% CI: 22.2–37.8 months) for glioblastomas. The median OS after the second operation was 14.1 months (95% CI: 7.7–20.5). In primary gliomas, 18 cases were low-grade gliomas, and 24 cases were high-grade gliomas. In recurrent gliomas, 34 cases were high-grade gliomas, and 8 cases were low-grade gliomas. Ten cases deteriorated from low-grade to a high-grade glioma (Table 1).

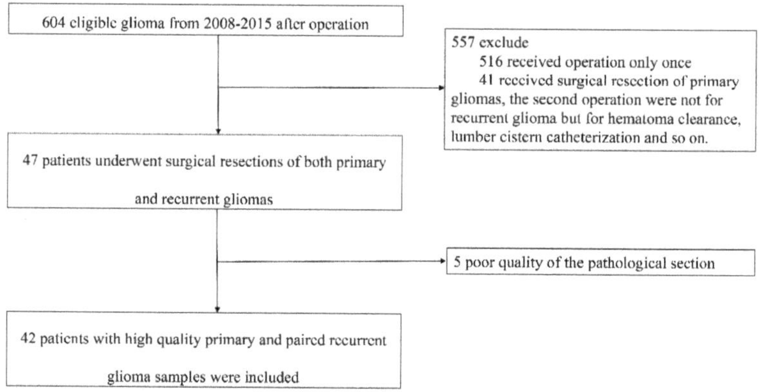

Figure 1. Study profile. Patients with high-quality primary and paired recurrent glioma samples were included.

Table 1. Baseline clinical characteristics.

Characteristics	Primary (*n* = 42)	Recurrence (*n* = 42)
Age(year)		
Mean (range)	43.2 (11–61)	44.9 (12–64)
Sex		
Male	27	27
Female	15	15
Tumor location—side		
Left	22	19
Right	20	22
Middle	0	1

Table 1. *Cont.*

Characteristics	Primary (*n* = 42)	Recurrence (*n* = 42)
Pathological Type		
Pilocytic astrocytoma	2	1
Astrocytoma, IDH1-mutant	8	9
Astrocytoma, IDH1-wildtype	15	10
Oligodendroglioma, NOS	3	4
Glioblastoma, IDH1-wildtype	12	18
Ganglioglioma	1	0
Pleomorphic xanthoastrocytoma	1	0
IDH1 status	9 (mutant)/33 (wildtype)	11 (mutant)/31 (wildtype)
WHO grade		
1	2	1
2	16	7
3	12	11
4	12	23
Residual after first surgery		
Yes	21	31
No	21	11
Adjuvant therapy after first surgery		
Radiotherapy alone	6	1
Chemotherapy alone	1	8
chemoradiotherapy	15	9
No adjuvant therapy	20	24
Type of recurrence		
Recurrence in situ		38
Distant relapse in the brain		4
Median PFS (month)	13.5	
Median OS (month)	33.8	14.1

3.2. Increased Protein Expression of PD-1 and PD-L1 in Recurrent Gliomas Compared to Their Corresponding Primary Tumors

In primary gliomas, expression of *PD-1* and *PD-L1* was found in 9 (22.0%) and 14 (33.3%) patients, respectively; in contrast, 25 (61.0%) and 31 (74.0%) of the lesions, respectively, had positive *PD-1* and *PD-L1* expression in their corresponding recurrent gliomas. In the samples from the first operation, *PD-1* and *PD-L1* expression was found in only 3 (16.7%) and 5 (27.8%), respectively, of patients with low-grade glioma; in high-grade glioma cases, the corresponding numbers of were 6 (26.1%) and 9 (37.5%), respectively. In the recurrent samples, low-grade gliomas expressed *PD-1* and *PD-L1* in 5 (62.5%) and 4 (50.0%), respectively, of the patients, whereas they were expressed in 21 (63.6%) and 26 (76.5%), respectively, of the patients with recurrent high-grade gliomas (Figure 2). For *PD-1* expression, 21 cases (51.2%) were shown to be negative in primary gliomas but positive in matched recurrent tumor samples; five cases (12.2%) were shown to be positive in primary gliomas but negative in matched recurrent tumor samples; 15 cases (36.6%) had no changes. The corresponding numbers for *PD-L1* were 24 (57.1%), 7 (16.7%), and 11 (26.2%), respectively. The overall expression rates of *PD-1* and *PD-L1* were enhanced after recurrence ($p < 0.005$; $p < 0.005$).

Ten low-grade glioma patients progressed to high-grade glioma: For *PD-1* expression, one patient showed positive expression in both the primary and recurrent samples; another patient changed from positive to negative. In the rest of the eight patients with *PD-1*-negative expression in primary glioma, three of them (3/8, 37.5%) changed to positive when there was recurrence. For *PD-L1* expression, one patient showed positive expression in both the primary and recurrent samples; two patients changed from positive to negative. In the rest of the seven patients with *PD-L1*-negative expression in primary glioma, five of them (5/7, 71.4%) changed to positive when there was recurrence.

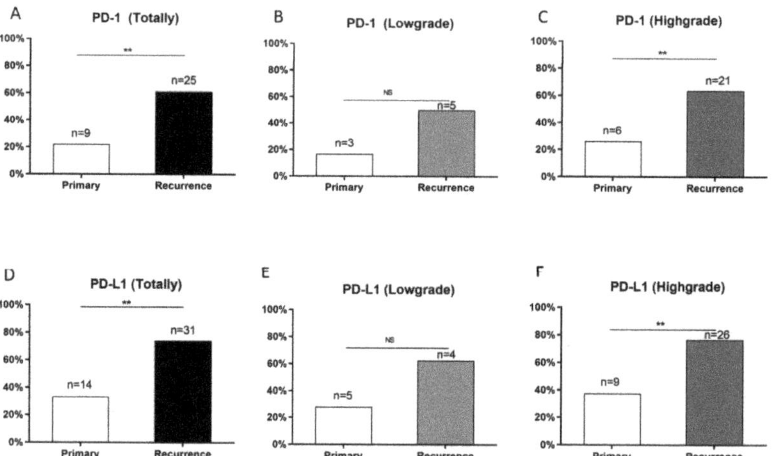

Figure 2. Expression of *PD-1* and *PD-L1* in primary and recurrent gliomas. (**A–C**) *PD-1* expression in primary and recurrent gliomas in the cohort of totality, low-grade and high-grade gliomas. (**D–F**) *PD-L1* expression in primary and recurrent gliomas in the cohort of totality, low-grade and high-grade gliomas. NS: no significance. **: $p < 0.01$.

In cases which were high-grade gliomas at first operation, they are still high-grade when there is recurrence. For *PD-1* expression, three patients showed positive expression in both the primary and recurrent samples; three patients changed from positive to negative. In the rest of the 17 patients with *PD-1*-negative expression in primary glioma, 14 of them (14/17, 82.4%) changed to positive when there was recurrence. For *PD-L1* expression, six patients showed positive expression in both the primary and recurrent samples; three patients changed from positive to negative. In the rest of the 15 patients with *PD-1*-negative expression in primary glioma, 14 of them (14/15, 93.3%) changed to positive when there was recurrence (Figure 3). Figure 4 showed the expression of *PD-1* and *PD-L1* in two typical patients in both their primary and recurrent gliomas.

3.3. Effect of Postoperative Adjuvant Therapy on the Expression of PD-1, PD-L1

To evaluate the impact of postoperative adjuvant treatment on the expression of *PD-1* and *PD-L1* genes, we further classified patients into the adjuvant therapy group ($n = 22$) and no adjuvant therapy group ($n = 20$). Compared with primary gliomas, the overall expression of *PD-1* increased in the no-adjuvant therapy group ($p < 0.05$), and it had an increasing trend in the adjuvant therapy group ($0.05 < p < 0.1$). Compared with primary gliomas, the overall expression of *PD-L1* increased in the adjuvant therapy group ($p < 0.005$), but the no adjuvant therapy group showed no statistical differences ($p > 0.25$) (Table 2).

Table 2. Effect of postoperative adjuvant therapy on the expression of *PD-1* and *PD-L1*.

	Group (*n*)	Primary (%)		Recurrence (%)		
		Positive	Negative	Positive	Negative	*p*
PD-1	No adjuvant therapy (20)	3	17	10	10	$p < 0.05$
	Adjuvant therapy (21)	6	15	15	6	$0.05 < p < 0.1$
PD-L1	No adjuvant therapy (20)	8	12	13	7	$p > 0.25$
	Adjuvant therapy (22)	6	16	18	4	$p < 0.005$

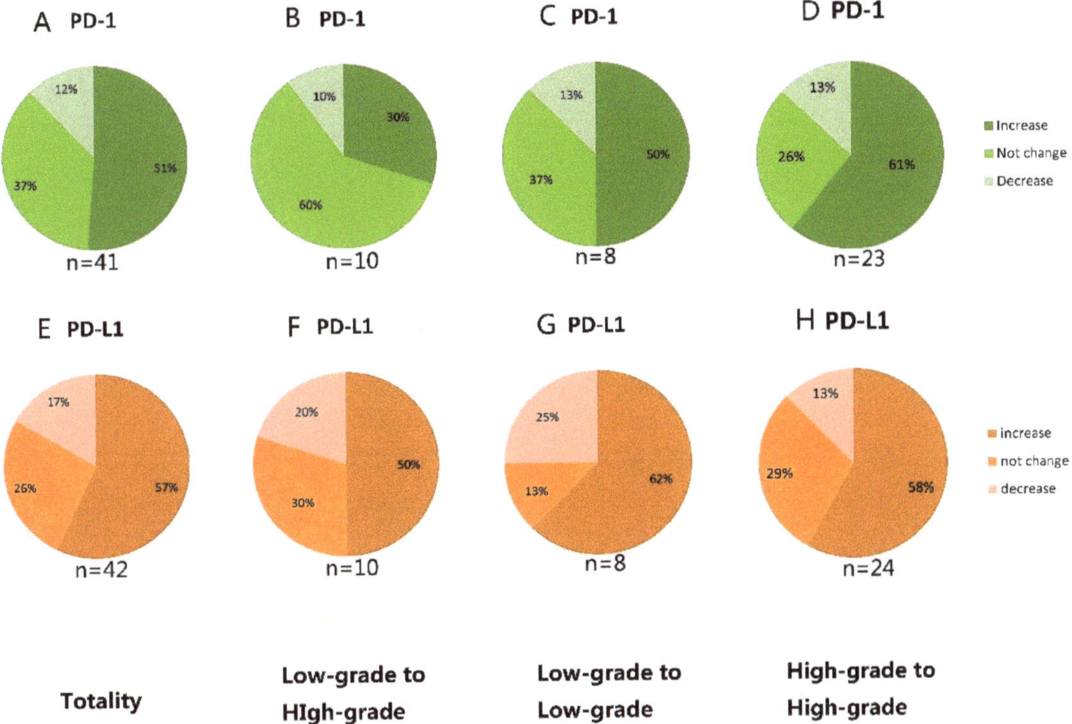

Figure 3. Change in *PD-1* (**A–D**) and *PD-L1* (**E–H**) expression from primary to recurrent glioma. (**A**) In the whole group, 51% of cases with *PD-1* expression increased from primary to recurrent glioma, 37% with *PD-1* expression were unchanged, and 12% with *PD-1* expression decreased. (**B**) Ten cases were low-grade gliomas at the first operation, and developed to high-grade when there was recurrence. Thirty percent of cases with *PD-1* expression increased from primary to recurrent glioma, 60% with *PD-1* expression were unchanged, and 13% with *PD-1* expression decreased. (**C**) Eight cases were low-grade gliomas at the first operation, and continued to be of low grade when there was recurrence. Fifty percent of cases with *PD-1* expression increased from primary to recurrent glioma, 37% with *PD-1* expression were unchanged, and 13% with *PD-1* expression decreased. (**D**) Twenty-three cases were high-grade gliomas at first operation, and were still high-grade gliomas when there was recurrence. Sixty-one percent of cases with *PD-1* expression increased from primary to recurrent glioma, 26% with *PD-1* expression were unchanged, and 13% with *PD-1* expression decreased. (**E**) In the whole group, 57% of cases with *PD-L1* expression increased from primary to recurrent glioma, 26% with *PD-L1* expression were unchanged, and 17% with *PD-L1* expression decreased. (**F**) Ten cases were low-grade gliomas at first operation and developed to high-grade when recurrence. Fifty percent of cases with *PD-L1* expression increased from primary to recurrent glioma, 30% with *PD-L1* expression were unchanged, and 20% with *PD-L1* expression decreased. (**G**) Eight cases were low-grade gliomas at the first operation and continued to be low-grade when there was recurrence. Sixty-two percent of cases with *PD-L1* expression increased from primary to recurrent glioma, 13% with *PD-L1* expression did not change, and 25% with *PD-L1* expression decreased. (**H**) Twenty-four cases were high-grade gliomas at first operation, and continued to be high-grade gliomas when there was recurrence. Fifty-eight percent of cases with *PD-L1* expression increased from primary to recurrent glioma, 29% with *PD-L1* expression were unchanged, and 13% with *PD-L1* expression decreased.

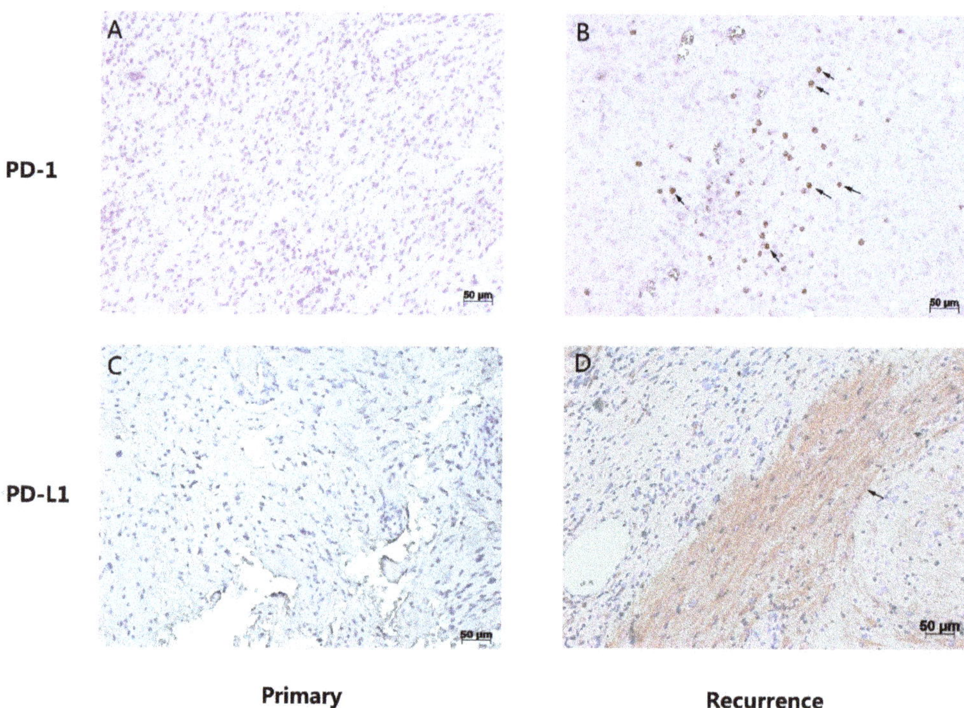

Figure 4. Expression of *PD-1* and *PD-L1* in two representative patients with primary and paired recurrent gliomas. One patient was diagnosed as having *PD-1*-negative glioblastoma after the first operation in 2013, and the histological diagnosis was still glioblastoma (**A**), with recurrence in 2014, but where the expression of *PD-1* was positive (**B**). Another patient was diagnosed as having *PD-L1*-negative anaplastic astrocytoma after the first operation in 2012 (**C**), and the histological diagnosis was glioblastoma when there was recurrence in 2014, and the expression of *PD-L1* was positive (**D**).

3.4. Correlation between PD-1/PD-L1 Expression

Recent studies indicated that co-expression of *PD-1* and *PD-L1* might be correlated with responses to immune checkpoint inhibitors [23]. Thus, we investigate the correlation between *PD-1/PD-L1* expression in glioma samples. In primary samples, co-expression of *PD-1* and *PD-L1* was evident in 5 (11.9%) glioma samples and negative staining for both *PD-1* and *PD-L1* in 23 (54.8%) samples. *PD-1* expression was not correlated with *PD-L1* status (rs = 0.198, p = 0.21) (Figure 5A). Notably, in recurrent samples, co-expression of *PD-1* and *PD-L1* was found in 23 (56.1%) samples and negative staining for both *PD-1* and *PD-L1* in 9 (22.0%) samples. *PD-1* expression showed a significant positive correlation with *PD-L1* expression (rs = 0.531, p < 0.001) (Figure 5B). Figure 6 shows *PD-1* and *PD-L1* expression in the same location in a primary tumor and its corresponding recurrent glioma from one of the glioma patients. This typical case was diagnosed as glioma grade III after surgical resection of primary glioma in 2012 with both the *PD-1* and *PD-L1*-negative stains. Strikingly, when recurrent, the pathologic diagnosis was glioma grade IV after secondary surgery in 2014 with both *PD-1* and *PD-L1*-positive staining. Here, we used *CD43* as a lymphocyte marker to the located *PD-1*-positive cells. Results indicated that *PD-1* was expressed in tumor-infiltrating lymphocytes, whereas *PD-L1* was expressed in tumor cells near these lymphocytes (Figure 6).

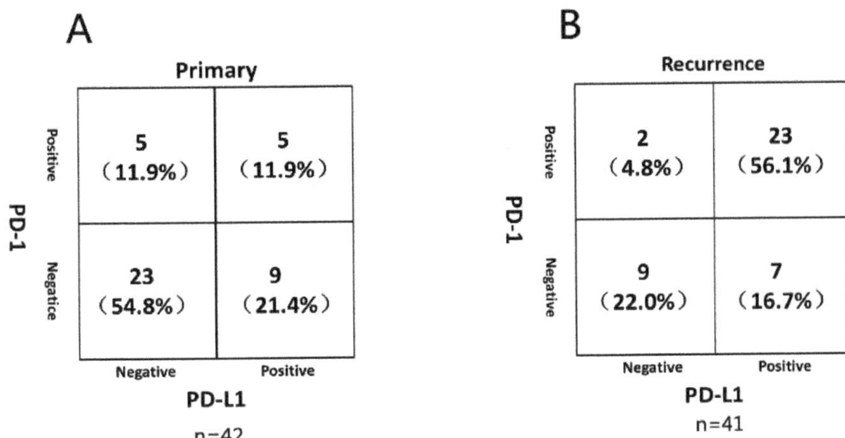

Figure 5. Correlation between *PD-1* and *PD-L1* in primary (**A**) and recurrent (**B**) gliomas.

3.5. The Prognostic Value of PD-1 and PD-L1

Table 3 compared baseline data of the *PD-1*-negative and *PD-1*-positive group. Table 4 showed baseline data of the *PD-L1*-negative and *PD-L1*-positive groups. Univariate analysis indicated that *PD-1* and *PD-L1* expression was not predictive of PFS or OS in primary and recurrent glioma. Importantly, for *PD-1*-negative primary gliomas, if the matched recurrent gliomas showed positive *PD-1* expression, the PFS became worse than the recurrent gliomas that remained *PD-1*-negative (12.7 vs. 25.9 months, $p = 0.032$). Interestingly, for *PD-L1*-negative primary gliomas, if the matched recurrent gliomas showed positive *PD-L1* expression, the OS became better than recurrent gliomas that remained *PD-1*-negative (33.8 vs. 17.5 months, $p < 0.001$) (Figure 7).

Table 3. Comparison of baseline clinical characteristics between *PD-1*-negative and *PD-1*-positive group in primary glioma.

Characteristics	*PD-1* Negative ($n = 32$)	*PD-1* Positive ($n = 10$)	*p*-Value
Age(year)			
Mean (range)	43 (11–61)	44 (12–61)	0.887
Sex			
Male	20	7	-
Female	12	3	1.000
WHO grade			
1	2	0	-
2	13	3	-
3	10	2	-
4	7	5	0.135
IDH1 status	9 (mutant)/23 (wildtype)	10 (wildtype)	0.086
Resection type			
Total gross resection	15	6	-
Subtotal resection	17	4	0.719
Adjuvant therapy after first surgery			
Yes	15	7	-
No	17	3	0.284
Type of recurrence			
Local recurrence	30	8	-
Distant recurrence	2	2	0.236
Median PFS (month)	15.3	10.3	0.081
Median OS (month)	34.0	24.4	0.183

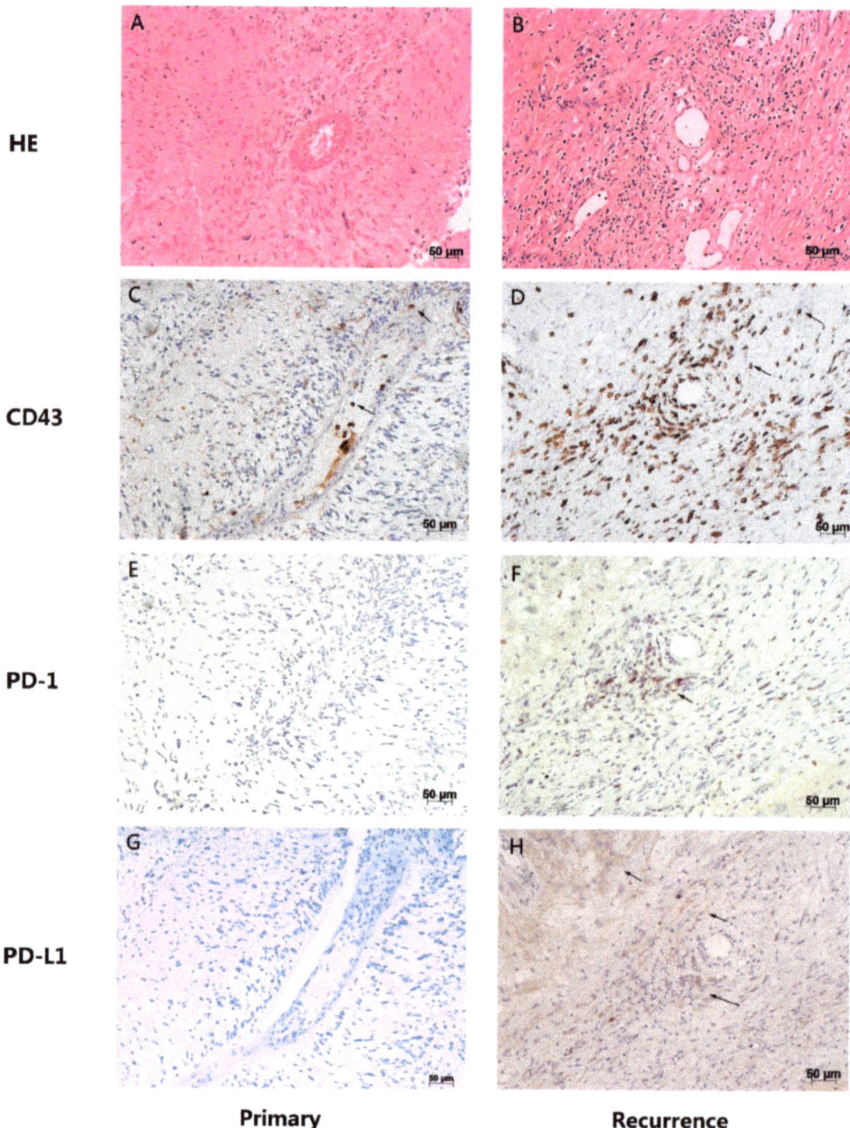

Primary   **Recurrence**

Figure 6. *PD-1* and *PD-L1* expression in the same location in a primary tumor and its corresponding recurrent glioma in one of the glioma patients. (**A**) HE staining for primary glioma. (**B**) HE stained for the corresponding recurrent glioma. (**C**) *CD43* immumohistochemical staining for the serial section of primary glioma. (**D**) *CD43* immumohistochemical staining for the serial section of the corresponding recurrent glioma. (**E**) *PD-1* immumohistochemical staining for the serial section of primary glioma. (**F**) *PD-1* immumohistochemical staining for the serial section of the corresponding recurrent glioma. (**G**) *PD-L1* immumohistochemical staining for the serial section of primary glioma. (**H**) *PD-L1* immumohistochemical staining for the serial section of the corresponding recurrent glioma. The patient was diagnosed as glioma grade III, negative for both *PD-1* and *PD-L1* after the first operation in 2011; when recurrence occurred, the histological diagnosis progressed to glioblastoma with positive *PD-1* and *PD-L1*. *PD-1* was expressed in lymphocytes (*CD43* labeled), and *PD-L1* was expressed in tumor cells near these lymphocytes.

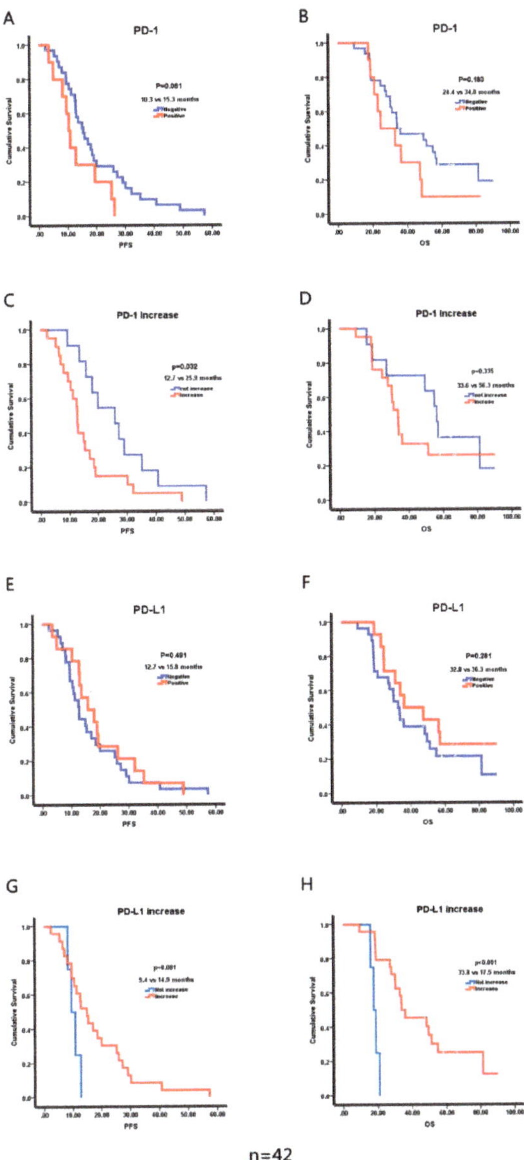

n=42

Figure 7. Correlation of the expression of *PD-1*, *PD-L1*, increasing *PD-1*, and increasing *PD-L1* with PFS and OS. (**A**) Comparation of progress-free survival (PFS) in *PD-1* negative group and *PD-1* positive group. (**B**) Comparation of overall survival (OS) in *PD-1* negative group and *PD-1* positive group. (**C**) Comparation of PFS in the group that *PD-1* expression didn't increase after recurrence and *PD-1* expression increased after recurrence. (**D**) Comparation of OS in the group that *PD-1* expression didn't increase after recurrence and *PD-1* expression increased after recurrence. (**E**) Comparation of PFS in *PD-L1* negative group and *PD-L1* positive group. (**F**) Comparation of OS in *PD-L1* negative group and *PD-L1* positive group. (**G**) Comparation of PFS in the group that *PD-L1* expression didn't increase after recurrence and *PD-L1* expression increased after recurrence. (**H**) Comparation of OS in the group that *PD-L1* expression didn't increase after recurrence and *PD-L1* expression increased after recurrence.

Table 4. Comparison of baseline clinical characteristics between *PD-L1*-negative and *PD-L1*-positive groups in primary glioma.

Characteristics	PD-L1 Negative (*n* = 28)	PD-L1 Positive (*n* = 14)	*p*-Value
Age(year)			
Mean (range)	46 (11–61)	38 (12–52)	0.073
Sex			
Male	17	10	-
Female	11	4	0.734
WHO grade			
1	2	0	-
2	11	5	-
3	7	5	-
4	8	4	0.573
IDH1 status	7 (mutant)/21 (wildtype)	2 (mutant)/12 (wildtype)	0.692
Resection type			
Total gross resection	16	5	-
Subtotal resection	12	9	0.326
Adjuvant therapy after first surgery			
Yes	16	6	-
No	12	8	0.515
Type of recurrence			
Local recurrence	25	13	-
Distant recurrence	3	1	1.000
Median PFS (month)	12.7	15.8	0.491
Median OS (month)	32.8	36.3	0.284

4. Discussion

PD signaling represents a tumor adaptation in the context of cancer. Tumor cells can restrain immune responses by endogenous cellular feedback, which is called "adaptive resistance". The combination of tumor PD-L1 and the PD-1 receptor on infiltrating effector T cells can inhibit cytotoxic activity. As a result, the CTL (cytotoxic T lymphocyte)-mediated elimination of cancer cells is inhibited (Figure 8). Elevated PD-L1 levels in tumor cells are connected to more aggressive behaviors and poor prognoses in certain cancers, including the breast, pancreas, kidney, ovary, gastric and esophageal cancer. To date, there is no systematic study on *PD-1/PD-L1* expression patterns in primary and paired recurrent glioma specimens. Berghoff et al. used immunohistochemistry to detect *PD-1/PD-L1* gene expression in 117 newly diagnosed glioblastomas and 18 recurrent glioblastomas. No significant difference in the expression of *PD-1* and diffuse/fibrillary *PD-L1* was found between the two groups. The expression of interspersed epithelioid tumor cells with membranous *PD-L1* in primary tumor cells was higher than that in recurrent specimens [26]. Rahman et al. analyzed 146 primary and 19 recurrent glioblastoma samples from TCGA datasets. They found no statistical difference in *PD-1* or *PD-L1* between primary and recurrent glioblastomas [27]. However, their studies have several limitations. The case number of recurrent gliomas in these studies is minimal, and they did not compare the paired primary and recurrent glioma, which cannot correctly answer the current question.

This study is the first to systematically compare the protein expression of the *PD-1/PD-L1* in primary and paired recurrent glioma tissue samples. Immunohistochemistry is currently the standard method to analyze the protein expression of *PD-1* and *PD-L1* in surgical specimens, which we used to detect the *PD-1/PD-L1* gene in 42 newly treated and paired recurrent glioma specimens. The expressions of *PD-1* and *PD-L1* in the recurrent samples were higher than those in the primary samples. As *PD-1* and *PD-L1* are immune escape-related genes, the increased expression of *PD-1* and *PD-L1* means that the tumor is more immune-tolerant. To investigate the effect of postoperative adjuvant therapy on the expression of *PD-1* and *PD-L1*, we further analyzed the cases in subgroups according to the situation of radiotherapy and chemotherapy. The results showed that in the chemoradiotherapy group, the expression of *PD-L1* in the recurrent specimens was significantly higher than that of the untreated samples, and *PD-1* had an increasing trend from primary to recurrence. The expression of *PD-1* was also significantly increased in the non-chemoradiotherapy group, but there was no significant difference in the expression of *PD-L1*. Our results suggest that the expression of *PD-1* in recurrent specimens is generally increased, which is not related to therapy, while the increase of *PD-L1* expression was more evident in patients who received adjuvant treatment. The expression rates of *PD-1* and *PD-L1* in primary glioma were 22% and 33%, respectively, and 61% and 74% in recurrent specimens. Berghoff et al. detected *PD-1/PD-L1* gene expression in 117 newly diagnosed glioblastomas and 18 recurrent glioblastomas. The expression rates of *PD-1* and *PD-L1* in untreated glioblastomas were 29.1% and 88%, respectively, and that of *PD-L1* in recurrent glioblastomas was 72.2%. Garber et al. analyzed *PD-1* protein expression in 235 glioma samples and *PD-L1* protein in 345 glioma samples and found that the expression rates of *PD-1* and *PD-L1* were 31.5% and 6.1%, respectively [24].

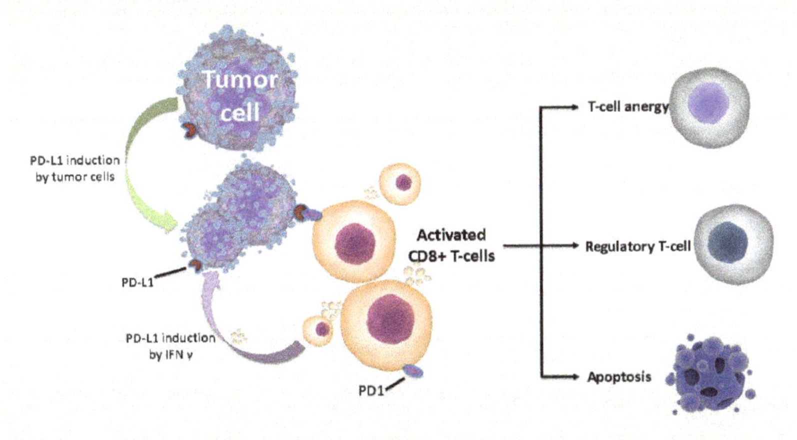

Figure 8. *PD-L1* is upregulated on cancer cells by IFN-γ released from infiltrating immune cells. Meanwhile, it is also upregulated by a tumor-specific non-IFN-γ-dependent mechanism. As a receptor of cancer cells, *PD-L1* induces a killing effect on cytotoxic T lymphocyte (CTL) by interacting with *PD-1* expressing tumor infiltrating lymphocyte (TIL) to inhibit the antitumor immune responses.

The expression rate of *PD-L1* was varied in different studies. Such variation might be caused by the different antibody used, the different scoring criteria, the heterogeneity of samples, and the different sample sizes. Berghoff et al. used the antibody made by Yale University Laboratory, the scoring standard was according to the staining range, and the specimen was glioblastoma [26]. Garber et al. used the antibody of Spring Biosciences and the scoring standard was according to the staining intensity, and the specimen was glioma [24]. However, we used the commercial antibody of cell signaling technology to identify glioma by staining range and staining intensity. We also found that *PD-1* and *PD-L1* expression were not correlated in primary glioma (Figure 4), somewhat at odds with the results reported by Garber et al. [24]. These findings may suggest that the *PD-1/PD-L1* pathway is activated to some extent in recurrent glioma, which means more severe immune suppression in recurrent glioma. Thus, the anti-*PD-1* and *PD-L1* antibodies used in glioma, especially in recurrent glioma, may be helpful.

We further studied the correlation between *PD-1*, *PD-L1* gene expression, and survival, and found that the expression of *PD-1* and *PD-L1* was not correlated with PFS and OS. Berghoff et al. analyzed the expression of the *PD-L1* protein in 446 glioma samples from the TCGA database by univariate analysis and multivariate analysis, and found that the expression of *PD-L1* was not related to overall survival. Besides, Nduom et al. analyzed *PD-1* and *PD-L1* mRNA expression by univariate analysis and found that they were associated with poor prognosis [28]. Hao et al. conducted a meta-analysis and found *PD-L1* high expression predicted poor prognosis in glioblastoma [29]. However, we found that patients in which *PD-1* expression increased from primary to recurrent glioma relapsed more efficiently than those whose *PD-1* expression did not increase. More interestingly, an increase in *PD-L1* from primary glioma to recurrent glioma was a better predictor of prognosis than was a lack of increase. As we know, *PD-1* prevents the activation of T-cell cytotoxicity, with or without interaction with its ligand. Therefore, cytotoxic T-cell activation is further suppressed in recurrent glioma. We conjecture that in the group whose *PD-L1* did not increase, a more effective tumor-promotion mechanism than *PD-L1* occurred in their recurrent glioma. This finding may partly explain the explosive growth of gliomas after a stationary phase achieved using an immune checkpoint inhibitor. However, in our study, only four patients were *PD-L1*-negative in both their primary and recurrent gliomas; we need a larger sample size to corroborate this hypothesis.

In recent years, immunosuppressive agents have shown considerable prospects in clinical practice. They are highly reactive in bladder cancer, head and neck cancer, melanoma, lung cancer, and renal cell carcinoma with persistent tumor remission [8,9]. The antibody targeting *PD-1* and *PD-L1* can reverse the immunosuppression, thus playing an anti-tumor effect [30]. An antibody directed to the *PD-1 / PD-L1* axis has shown a substantial impact on the mouse glioma model [31]. Currently, many clinical trials are underway on the use of these antibodies in gliomas—a phase III randomized controlled clinical trial combined nivolumab and ipilimumab in the treatment of glioblastoma (NCT02017717). A phase II clinical trial applied pembrolizumab alone or combined with bevacizumab to treat recurrent glioblastoma (NCT02337491). A phase II clinical trial combined atezolizumab with radiation and temozolomide during the concurrent stage and in combination with temozolomide during the adjuvant stage in newly diagnosed glioblastoma (NCT03174197), while Checkmate 143 reported that nivolumab, which targets *PD-1*, did not show survival benefits compared with bevacizumab in recurrent glioblastoma [32]. Possible etiologies of treatment failure may be systemic lymphopenia when recurrence, low *PD-L1* expression rate in this cohort, poor drug penetration of the blood–brain barrier (BBB), antigen-specific T-cells heavily dysfunctional in this special immune microenvironment, no identification of discrepancies between different genomic subtypes in their response to *PD-1/PD-L1* checkpoint blockades, or the presence of structural barriers preventing T-cell–antibody interactions. Therefore, two other studies followed. However, the Phase III CheckMate-498 study did not meet the primary endpoint of OS with nivolumab plus radiation in patients with newly diagnosed MGMT-unmethylated glioblastoma. CheckMate-548, which compared nivolumab or placebo with radiotherapy plus temozolomide (TMZ) in patients with newly diagnosed glioblastoma with a methylated *MGMT* promotor, obtained a negative result. Although these results are disappointing, many new methods to solve these potential causes of failure and new explorations are ongoing. A team from the Department of Immunobiology, Yale University School of Medicine found that the existence of GBM in the brain alone is not enough to cause an immune antitumor effect. The activation of a peripheral immune response may help to prolong the survival of patients with GBM. Their established VEGF-C-mRNA (lymphangiogenesis-promoting factor, Vascular Endothelial Growth Factor C (VEGF-C), using AAV9 and mRNA delivery vectors) can increase the activation of tumor-specific T cells and their tumor infiltration in lymph nodes. Combined administration of VEGF-C-mRNA and the anti-*PD-1* antibody resulted in tumor regression and survival benefits in tumor-bearing mice in a T-cell-dependent manner [33]. Hwang et al. reported that laser interstitial thermotherapy (LITT) can destroy BBB and may enhance host T-cell-mediated cytotoxicity. The combination of LITT and pembrolizumab in recurrent IDH wild-type glioblastoma prolonged PFS and OS [34,35]. A phase II clinical trial NCT03661723 adopted reirradiation-combined Pembrolizumab to stimulate the immune response and open up the blood–brain barrier, and achieved further stratification with Bevacizumab

Naïve and Bevacizumab Resistant Recurrent Glioblastoma to accurately capture the group benefiting from *PD-1* antibody-combined irradiation [36]. There are also similar studies, such as NCT03743662 [37], NCT03197506 [38], NCT04977375 [39]. NCT03233152 adopted intra-tumoral ipilimumab plus intravenous nivolumab following the resection of recurrent glioblastoma for circumvention of BBB and stronger immune activation [40]. NCT04013672 used SurVaxM (a survivin vaccine) plus pembrolizumab for recurrent glioblastoma [41]. A single-arm phase II clinical trial (NCT02550249) concluded that neoadjuvant nivolumab modifies the tumor immune status in resectable glioblastoma [42]. A randomized, multi-institution clinical trial conducted by the Ivy Foundation Early-Phase Clinical Trials Consortium demonstrated that neoadjuvant anti-*PD-1* immunotherapy promotes a survival benefit with intratumoral and systemic immune responses compared to the adjuvant-only group in recurrent glioblastoma [43].

Current data are not mature and early studies have not shown clear benefits, yet we cannot rule out the possibility of using immune checkpoint inhibitors as a potential strategy for glioma. Our results indicated that the protein expression of *PD-1* and *PD-L1* increased in recurrent gliomas. Several clinical trials have demonstrated that the expression of *PD-1* and *PD-L1* is correlated with treatment response [9,44]. Immune checkpoint inhibitors targeting *PD-1* and *PD-L1* may have potential value in the treatment of recurrent gliomas.

5. Conclusions

This study explored the *PD-1/PD-L1* protein expression in recurrent glioma and its paired primary tumor. We confirmed the tendency of increased *PD-1/PD-L1* in recurrent glioma. In addition, increased *PD-L1* expression was related to adjuvant treatment. We revealed some immune characteristics of primary and their paired recurrent gliomas to a better understanding of gliomas, which indicated the potential therapeutic and predictive value of *PD-1* and *PD-L1* in the treatment and diagnosis of recurrent gliomas. Although several large-scale clinical trials obtained negative results of *PD-1* antibody applied in glioblastoma, further studies are needed to recognize the special immune microenvironment of the brain, especially under the tumor condition. We believe the dawn is just around the corner.

Author Contributions: Q.W. made designed the study and determined the final version. W.Y. drafted the manuscript, did the experiment and made the figures and tables. A.S. helped to make the figures and polished the article. X.R. did part of the experiment. Z.C. conducted the statistic analysis. J.X. help to read the immunohistochemical sections. All authors have read and agreed to the published version of the manuscript.

Funding: This work was supported by the National Natural Science Foundation of China (82073332).

Institutional Review Board Statement: The study was conducted according to the guidelines of the Declaration of Helsinki, and approved by the Institutional Ethics Committee of SAHZU (protocol code: 046).

Informed Consent Statement: Informed consent was obtained from all subjects involved in the study.

Data Availability Statement: Data sharing not applicable to this article as no datasets were generated or analyzed during the current study.

Conflicts of Interest: The authors declare that they have no conflict of interest.

References

1. Kent, W.J. BLAT–the BLAST-like alignment tool. *Genome Res.* **2002**, *12*, 656–664. [PubMed]
2. Pruitt, K.D.; Tatusova, T.; Maglott, D.R. NCBI Reference Sequence (RefSeq): A curated non-redundant sequence database of genomes, transcripts and proteins. *Nucleic Acids Res.* **2005**, *33*, D501–D504. [CrossRef] [PubMed]
3. Wang, X.; Guo, G.; Guan, H.; Yu, Y.; Lu, J.; Yu, J. Challenges and potential of PD-1/PD-L1 checkpoint blockade immunotherapy for glioblastoma. *J. Exp. Clin. Cancer Res.* **2019**, *38*, 87. [CrossRef] [PubMed]
4. Stupp, R.; Mason, W.P.; van den Bent, M.J.; Weller, M.; Fisher, B.; Taphoorn, M.J.; Belanger, K.; Brandes, A.A.; Marosi, C.; Bogdahn, U.; et al. Radiotherapy plus concomitant and adjuvant temozolomide for glioblastoma. *N. Engl. J. Med.* **2005**, *352*, 987–996. [CrossRef] [PubMed]

5. Caccese, M.; Indraccolo, S.; Zagonel, V.; Lombardi, G. PD-1/PD-L1 immune-checkpoint inhibitors in glioblastoma: A concise review. *Crit. Rev. Oncol. Hematol.* **2019**, *135*, 128–134. [CrossRef]

6. Litak, J.; Mazurek, M.; Grochowski, C.; Kamieniak, P.; Rolinski, J. PD-L1/PD-1 Axis in Glioblastoma Multiforme. *Int. J. Mol. Sci* **2019**, *20*, 5347. [CrossRef]

7. Tatter, S.B. Recurrent malignant glioma in adults. *Curr. Treat. Options Oncol.* **2002**, *3*, 509–524. [CrossRef]

8. Wolchok, J.D.; Kluger, H.; Callahan, M.K.; Postow, M.A.; Rizvi, N.A.; Lesokhin, A.M.; Segal, N.H.; Ariyan, C.E.; Gordon, R.A.; Reed, K.; et al. Nivolumab plus ipilimumab in advanced melanoma. *N. Engl. J. Med.* **2013**, *369*, 122–133. [CrossRef]

9. Topalian, S.L.; Hodi, F.S.; Brahmer, J.R.; Gettinger, S.N.; Smith, D.C.; McDermott, D.F.; Powderly, J.D.; Carvajal, R.D.; Sosman, J.A.; Atkins, M.B.; et al. Safety, activity, and immune correlates of anti-PD-1 antibody in cancer. *N. Engl. J. Med.* **2012**, *366*, 2443–2454. [CrossRef]

10. Goldman, J.W.; Dvorkin, M.; Chen, Y.; Reinmuth, N.; Hotta, K.; Trukhin, D.; Statsenko, G.; Hochmair, M.J.; Ozguroglu, M.; Ji, J.H.; et al. Durvalumab, with or without tremelimumab, plus platinum-etoposide versus platinum-etoposide alone in first-line treatment of extensive-stage small-cell lung cancer (CASPIAN): Updated results from a randomised, controlled, open-label, phase 3 trial. *Lancet Oncol.* **2021**, *22*, 51–65. [CrossRef]

11. Waterhouse, D.M.; Garon, E.B.; Chandler, J.; McCleod, M.; Hussein, M.; Jotte, R.; Horn, L.; Daniel, D.B.; Keogh, G.; Creelan, B.; et al. Continuous Versus 1-Year Fixed-Duration Nivolumab in Previously Treated Advanced Non-Small-Cell Lung Cancer: CheckMate 153. *J. Clin. Oncol.* **2020**, *38*, 3863–3873. [CrossRef] [PubMed]

12. Jacobs, J.F.; Idema, A.J.; Bol, K.F.; Nierkens, S.; Grauer, O.M.; Wesseling, P.; Grotenhuis, J.A.; Hoogerbrugge, P.M.; de Vries, I.J.; Adema, G.J. Regulatory T cells and the PD-L1/PD-1 pathway mediate immune suppression in malignant human brain tumors. *Neuro-Oncology* **2009**, *11*, 394–402. [CrossRef] [PubMed]

13. Dunn, G.P.; Dunn, I.F.; Curry, W.T. Focus on TILs: Pro.ognostic significance of tumor infiltrating lymphocytes in human glioma. *Cancer Immun.* **2007**, *7*, 12. [PubMed]

14. Croci, D.O.; Zacarias Fluck, M.F.; Rico, M.J.; Matar, P.; Rabinovich, G.A.; Scharovsky, O.G. Dynamic cross-talk between tumor and immune cells in orchestrating the immunosuppressive network at the tumor microenvironment. *Cancer Immunol. Immunother.* **2007**, *56*, 1687–1700. [CrossRef] [PubMed]

15. Blank, C.; Gajewski, T.F.; Mackensen, A. Interaction of PD-L1 on tumor cells with PD-1 on tumor-specific T cells as a mechanism of immune evasion: Implications for tumor immunotherapy. *Cancer Immunol. Immunother.* **2005**, *54*, 307–314. [CrossRef]

16. Freeman, G.J.; Long, A.J.; Iwai, Y.; Bourque, K.; Chernova, T.; Nishimura, H.; Fitz, L.J.; Malenkovich, N.; Okazaki, T.; Byrne, M.C.; et al. Engagement of the PD-1 immunoinhibitory receptor by a novel B7 family member leads to negative regulation of lymphocyte activation. *J. Exp. Med.* **2000**, *192*, 1027–1034. [CrossRef]

17. Martner, A.; Thoren, F.B.; Aurelius, J.; Hellstrand, K. Immunotherapeutic strategies for relapse control in acute myeloid leukemia. *Blood Rev.* **2013**, *27*, 209–216. [CrossRef]

18. Bekiaris, V.; Sedy, J.R.; Macauley, M.G.; Rhode-Kurnow, A.; Ware, C.F. The inhibitory receptor BTLA controls gammadelta T cell homeostasis and inflammatory responses. *Immunity* **2013**, *39*, 1082–1094. [CrossRef]

19. Hebeisen, M.; Baitsch, L.; Presotto, D.; Baumgaertner, P.; Romero, P.; Michielin, O.; Speiser, D.E.; Rufer, N. SHP-1 phosphatase activity counteracts increased T cell receptor affinity. *J. Clin. Investig.* **2013**, *123*, 1044–1056. [CrossRef]

20. Taube, J.M.; Klein, A.; Brahmer, J.R.; Xu, H.; Pan, X.; Kim, J.H.; Chen, L.; Pardoll, D.M.; Topalian, S.L.; Anders, R.A. Association of PD-1, PD-1 ligands, and other features of the tumor immune microenvironment with response to anti-PD-1 therapy. *Clin. Cancer Res.* **2014**, *20*, 5064–5074. [CrossRef]

21. Bloch, O.; Crane, C.A.; Kaur, R.; Safaee, M.; Rutkowski, M.J.; Parsa, A.T. Gliomas promote immunosuppression through induction of B7-H1 expression in tumor-associated macrophages. *Clin. Cancer Res.* **2013**, *19*, 3165–3175. [CrossRef] [PubMed]

22. Liu, Y.; Carlsson, R.; Ambjorn, M.; Hasan, M.; Badn, W.; Darabi, A.; Siesjo, P.; Issazadeh-Navikas, S. PD-L1 expression by neurons nearby tumors indicates better prognosis in glioblastoma patients. *J. Neurosci.* **2013**, *33*, 14231–14245. [CrossRef] [PubMed]

23. Zeng, J.; See, A.P.; Phallen, J.; Jackson, C.M.; Belcaid, Z.; Ruzevick, J.; Durham, N.; Meyer, C.; Harris, T.J.; Albesiano, E.; et al. Anti-PD-1 blockade and stereotactic radiation produce long-term survival in mice with intracranial gliomas. *Int. J. Radiat. Oncol. Biol. Phys.* **2013**, *86*, 343–349. [CrossRef] [PubMed]

24. Garber, S.T.; Hashimoto, Y.; Weathers, S.P.; Xiu, J.; Gatalica, Z.; Verhaak, R.G.; Zhou, S.; Fuller, G.N.; Khasraw, M.; de Groot, J.; et al. Immune checkpoint blockade as a potential therapeutic target: Surveying CNS malignancies. *Neuro-Oncology* **2016**, *18*, 1357–1366. [CrossRef]

25. Gatalica, Z.; Snyder, C.; Maney, T.; Ghazalpour, A.; Holterman, D.A.; Xiao, N.; Overberg, P.; Rose, I.; Basu, G.D.; Vranic, S.; et al. Programmed cell death 1 (PD-1) and its ligand (PD-L1) in common cancers and their correlation with molecular cancer type. *Cancer Epidemiol. Biomarkers Prev.* **2014**, *23*, 2965–2970. [CrossRef]

26. Berghoff, A.S.; Kiesel, B.; Widhalm, G.; Rajky, O.; Ricken, G.; Wohrer, A.; Dieckmann, K.; Filipits, M.; Brandstetter, A.; Weller, M.; et al. Programmed death ligand 1 expression and tumor-infiltrating lymphocytes in glioblastoma. *Neuro-Oncology* **2015**, *17*, 1064–1075. [CrossRef]

27. Rahman, M.; Kresak, J.; Yang, C.; Huang, J.; Hiser, W.; Kubilis, P.; Mitchell, D. Analysis of immunobiologic markers in primary and recurrent glioblastoma. *J. Neurooncol.* **2018**, *137*, 249–257. [CrossRef]

28. Nduom, E.K.; Wei, J.; Yaghi, N.K.; Huang, N.; Kong, L.Y.; Gabrusiewicz, K.; Ling, X.; Zhou, S.; Ivan, C.; Chen, J.Q.; et al. PD-L1 expression and prognostic impact in glioblastoma. *Neuro-Oncology* **2016**, *18*, 195–205. [CrossRef]

29. Hao, C.; Chen, G.; Zhao, H.; Li, Y.; Chen, J.; Zhang, H.; Li, S.; Zhao, Y.; Chen, F.; Li, W.; et al. PD-L1 Expression in Glioblastoma, the Clinical and Prognostic Significance: A Systematic Literature Review and Meta-Analysis. *Front. Oncol.* **2020**, *10*, 1015. [CrossRef]
30. Hirano, F.; Kaneko, K.; Tamura, H.; Dong, H.; Wang, S.; Ichikawa, M.; Rietz, C.; Flies, D.B.; Lau, J.S.; Zhu, G.; et al. Blockade of B7-H1 and PD-1 by monoclonal antibodies potentiates cancer therapeutic immunity. *Cancer Res.* **2005**, *65*, 1089–1096.
31. Wainwright, D.A.; Chang, A.L.; Dey, M.; Balyasnikova, I.V.; Kim, C.K.; Tobias, A.; Cheng, Y.; Kim, J.W.; Qiao, J.; Zhang, L.; et al. Durable therapeutic efficacy utilizing combinatorial blockade against IDO, CTLA-4, and PD-L1 in mice with brain tumors. *Clin. Cancer Res.* **2014**, *20*, 5290–5301. [CrossRef] [PubMed]
32. Filley, A.C.; Henriquez, M.; Dey, M. Recurrent glioma clinical trial, CheckMate-143: The game is not over yet. *Oncotarget* **2017**, *8*, 91779–91794. [CrossRef] [PubMed]
33. Song, E.; Mao, T.; Dong, H.; Boisserand, L.S.B.; Antila, S.; Bosenberg, M.; Alitalo, K.; Thomas, J.L.; Iwasaki, A. VEGF-C-driven lymphatic drainage enables immunosurveillance of brain tumours. *Nature* **2020**, *577*, 689–694. [CrossRef] [PubMed]
34. Campian, J.; Butt, O.; Ghinaseddin, A.; Rahman, M.; Chheda, M.; Johanns, T.; Ansstas, G.; Huang, J.; Liu, J.; Talcott, G.; et al. CTIM-26. PHASE I/II STUDY OF THE COMBINATION OF PEMBROLIZUMAB (MK-3475) AND LASER INTERSTITIAL THERMAL THERAPY (LITT) IN RECURRENT GLIOBLASTOMA. *Neuro-Oncology* **2021**, *23*, vi56. [CrossRef]
35. Hwang, H.; Huang, J.; Khaddour, K.; Butt, O.H.; Ansstas, G.; Chen, J.; Katumba, R.G.; Kim, A.H.; Leuthardt, E.C.; Campian, J.L. Prolonged response of recurrent IDH-wild-type glioblastoma to laser interstitial thermal therapy with pembrolizumab. *CNS Oncol.* **2022**, CNS81. [CrossRef] [PubMed]
36. Pembrolizumab and Reirradiation in Bevacizumab Naïve and Bevacizumab Resistant Recurrent Glioblastoma. Available online: https://clinicaltrials.gov/ct2/show/NCT03661723?term=NCT03661723&draw=2&rank=1 (accessed on 7 September 2018).
37. Nivolumab With Radiation Therapy and Bevacizumab for Recurrent MGMT Methylated Glioblastoma. Available online: https://clinicaltrials.gov/ct2/show/NCT03743662?term=NCT03743662&draw=2&rank=1 (accessed on 16 November 2018).
38. Pembrolizumab and Standard Therapy in Treating Patients With Glioblastoma. Available online: https://clinicaltrials.gov/ct2/show/NCT03197506?term=NCT03197506&draw=2&rank=1 (accessed on 23 June 2017).
39. Trial of Anti-PD-1 Immunotherapy and Stereotactic Radiation in Patients With Recurrent Glioblastoma. Available online: https://clinicaltrials.gov/ct2/show/NCT04977375?term=NCT04977375&draw=2&rank=1 (accessed on 26 July 2021).
40. Intra-tumoral Ipilimumab Plus Intravenous Nivolumab Following the Resection of Recurrent Glioblastoma (GlitIpNi). Available online: https://clinicaltrials.gov/ct2/show/NCT03233152?term=NCT03233152&draw=2&rank=1 (accessed on 28 July 2017).
41. Fenstermaker, R.A.; Ciesielski, M.J. Challenges in the development of a survivin vaccine (SurVaxM) for malignant glioma. *Expert Rev. Vaccines* **2014**, *13*, 377–385. [CrossRef]
42. Schalper, K.A.; Rodriguez-Ruiz, M.E.; Diez-Valle, R.; Lopez-Janeiro, A.; Porciuncula, A.; Idoate, M.A.; Inoges, S.; de Andrea, C.; Lopez-Diaz de Cerio, A.; Tejada, S.; et al. Neoadjuvant nivolumab modifies the tumor immune microenvironment in resectable glioblastoma. *Nat. Med.* **2019**, *25*, 470–476. [CrossRef]
43. Cloughesy, T.F.; Mochizuki, A.Y.; Orpilla, J.R.; Hugo, W.; Lee, A.H.; Davidson, T.B.; Wang, A.C.; Ellingson, B.M.; Rytlewski, J.A.; Sanders, C.M.; et al. Neoadjuvant anti-PD-1 immunotherapy promotes a survival benefit with intratumoral and systemic immune responses in recurrent glioblastoma. *Nat. Med.* **2019**, *25*, 477–486. [CrossRef]
44. Tumeh, P.C.; Harview, C.L.; Yearley, J.H.; Shintaku, I.P.; Taylor, E.J.; Robert, L.; Chmielowski, B.; Spasic, M.; Henry, G.; Ciobanu, V.; et al. PD-1 blockade induces responses by inhibiting adaptive immune resistance. *Nature* **2014**, *515*, 568–571. [CrossRef]

Article

Identification of an IL-4-Related Gene Risk Signature for Malignancy, Prognosis and Immune Phenotype Prediction in Glioma

Ying Qi [1,2], Xinyu Yang [1,2], Chunxia Ji [1,2], Chao Tang [1,2] and Liqian Xie [1,2,*]

1   Department of Neurosurgery, Huashan Hospital, Shanghai Medical College, Fudan University, Shanghai 200040, China; 18211220054@fudan.edu.cn (Y.Q.); 20211220056@fudan.edu.cn (X.Y.); 13211010007@fudan.edu.cn (C.J.); chaotang@fudan.edu.cn (C.T.)
2   Neurosurgical Institute of Fudan University, Fudan University, Shanghai 200040, China
*   Correspondence: 071105227@fudan.edu.cn

Abstract: Background: Emerging molecular and genetic biomarkers have been introduced to classify gliomas in the past decades. Here, we introduced a risk signature based on the cellular response to the IL-4 gene set through Least Absolute Shrinkage and Selection Operator (LASSO) regression analysis. Methods: In this study, we provide a bioinformatic profiling of our risk signature for the malignancy, prognosis and immune phenotype of glioma. A cohort of 325 patients with whole genome RNA-seq expression data from the Chinese Glioma Genome Atlas (CGGA) dataset was used as the training set, while another cohort of 667 patients from The Cancer Genome Atlas (TCGA) dataset was used as the validating set. The LASSO model identified a 10-gene signature which was considered as the optimal model. Results: The signature was confirmed to be a good predictor of clinical and molecular features involved in the malignancy of gliomas. We also identified that our risk signature could serve as an independently prognostic biomarker in patients with gliomas ($p < 0.0001$). Correlation analysis showed that our risk signature was strongly correlated with the Tregs, M0 macrophages and NK cells infiltrated in the microenvironment of glioma, which might be a supplement to the existing incomplete innate immune mechanism of glioma phenotypes. Conclusions: Our IL-4-related gene signature was associated with more aggressive and immunosuppressive phenotypes of gliomas. The risk score could predict prognosis independently in glioma, which might provide a new insight for understanding the IL-4 involved mechanism of gliomas.

Keywords: glioma; IL-4; 10-gene signature; prognosis; microenvironment

Citation: Qi, Y.; Yang, X.; Ji, C.; Tang, C.; Xie, L. Identification of an IL-4-Related Gene Risk Signature for Malignancy, Prognosis and Immune Phenotype Prediction in Glioma. *Brain Sci.* **2022**, *12*, 181. https://doi.org/10.3390/brainsci12020181

Academic Editors: Lucia Lisi and Andrew Clarkson

Received: 5 January 2022
Accepted: 27 January 2022
Published: 29 January 2022

Publisher's Note: MDPI stays neutral with regard to jurisdictional claims in published maps and institutional affiliations.

1. Introduction

Gliomas are the most prevalent and aggressive brain tumors, with extremely poor prognosis in adults. Among all grades of gliomas, glioblastoma (GBM) is the most devastating type with a median overall survival time of approximately 19 months [1]. At recurrence, patients always have a very poor survival rate despite the beneficial treatments including second surgery [1] and re-irradiation [2]. In the past decade, newly emerging therapeutic approaches such as tumor-treating fields (TTF) and several immunotherapies were introduced in the hopes of GBM treatment [3]. However, the majority of the immunotherapies including PD-1/PD-L1 checkpoint inhibitors, chimeric antigen receptor-T cells (CAR-T), and adoptive T cell strategies ended in the failure of GBM treatments [4–6]. These failures strongly indicated that beyond the T cell-based adaptive immunity, innate immunity might be one of the most critical aspects to regulate anti-tumor immunity in the glioma microenvironment [4]. Many works have focused on the microglia, while other innate immune cells such as infiltrating macrophages and NK cells are becoming more attractive in the studies of the GBM immune microenvironment [7,8]. Despite the existing efforts in glioma research, little progress has been made in understanding the molecular mechanism of gliomas, and the effects of innate immunity in the glioma microenvironment still remain incomplete [9].

Interleukin-4 (IL-4), a Th2 cytokine mainly produced by activated T cells and mast cells, is confirmed to regulate the proliferation of lymphocytes [10]. In addition to its immune function, IL-4 produced by cancer cells is also reported to promote tumor proliferation and aggressiveness in glioma, bladder cancer, breast cancer and other epithelial tumors through STAT6 signal transduction pathways [11,12]. Enhanced IL-4 secretion of cancer cells could also be involved in tumor-associated macrophages (TAM)-induced tumor growth and metastasis [13]. Polymorphisms in the IL-4 receptor genes are reported to influence the glioma survival, which indicate that IL-4-induced inflammatory pathways might regulate the glioma development and prognosis [14]. However, the role of cellular response to IL-4 in glioma development remains unclear.

In our study, we focus on the cellular response to the IL-4 gene set from gene ontology in gliomas. Consensus clustering was firstly applied to identify that the cellular response to the IL-4 gene set had the ability to distinguish clinicopathological features of gliomas. Next, a cellular response to IL-4-related gene risk signature was generated in the CGGA dataset through LASSO regression, and further validated in the TCGA dataset. Our risk signature was observed to be strongly associated with clinicopathological features of gliomas and to be an independent prognostic factor for both all grade gliomas and GBM. In addition, we found that this cellular response to the IL-4-related gene risk signature was closely related to tumor infiltrating lymphocytes (including M0 macrophages, NK cells, and Tregs) in the glioma microenvironment, which indicated a potential association between the cellular response to IL-4 and the immune phenotype of gliomas. We believe that our bioinformatic analysis might provide a new insight for understanding the IL-4 involved mechanism of gliomas.

2. Materials and Methods

2.1. Data Collection

Two population datasets were analyzed in this study: a glioma dataset from the Chinese Glioma Genome Atlas (CGGA) dataset and a glioma dataset from The Cancer Genome Atlas (TCGA) dataset. The RNA-seq expression data and clinical information of 325 glioma patients from CGGA dataset (http://www.cgga.org.cn, 1 October 2020) was used as the training set [15,16]. RNA-seq data and clinical information of 667 glioma patients from TCGA dataset (http://cancergenome.nih.gov, 1 October 2020) was used as validation set [17,18].

2.2. Consensus Clustering

All 28 genes in the cellular response to IL-4-related gene set were extracted from Molecular Signatures Database v6.2 [15]. Determined by the median absolute deviation (MAD), the most variable genes of the cellular response to IL-4-related gene set were selected for the consensus clustering analysis through ConsensusClusterPlus package [19]. R programming language was used for consensus clustering for detecting the cellular response to IL-4-related glioma subgroups of the CGGA training set. The optimal number of the clusters was further determined by quantitative stability evidence in an unsupervised analysis.

2.3. Construction of the Gene Risk Signature

Screened by univariate Cox regression analysis in the CGGA training set, all genes with high prognostic value ($p < 0.05$) in the cellular response to IL-4-related gene set were selected for Least Absolute Shrinkage and Selection Operator (LASSO) regression analysis [20]. We used glmnet package in R programming language as our LASSO regression tool. The generalized linear model produced by LASSO regression analysis was further analyzed with 10-fold cross validation in order to generate the minimum cross validated error [21]. Based on the minimum cross validated error, expressions (expr) of 10 genes in the cellular response to IL-4-related gene set and their regression coefficients (Coef) were eventually achieved. Then, the risk score for each patient in the CGGA training set and TCGA validation set was calculated by the following formula: risk score = exprgene1 × Coefgene1 + exprgene2 × Coefgene2 + ... + exprgene10 × Coefgene10.

All patients in the CGGA training set and TCGA validation set were then separated into high or low risk group according to the median risk score cutoff. Survival analysis based on the risk score was evaluated by Kaplan–Meier survival curve using R programming language. Univariate and multivariate survival analysis was performed by using Cox proportional hazards model in R programming language.

2.4. Estimation of the Abundances of Immune Cell Types

For evaluating the association between the cellular response to IL-4-related gene risk signature and the immune phenotype of glioma, estimation of the abundances of immune cell types through gene expression data in CGGA and TCGA datasets was achieved by CIBERSORT package in R programming language. We used LM22 introduced by Newman, A.M. et al. as our input marker matrix of 22 types of immune cells [22]. The correlation between our risk signature and immune cell was validated by corrplot package in R programming language, and all the heatmaps were produced through ComplexHeatmap package in R programming language [23].

2.5. Statistical Analyses

Main statistical analysis including Student's *t*-test, chi-square test, and Pearson's test were also performed in R programming language. Statistical significance was considered at the level of $p < 0.05$.

3. Results

3.1. Classification of Gliomas Based on Cellular Response to IL-4-Related Gene Set

The gene expression profiling of all genes in the cellular response to IL-4-related gene set obtained from the CGGA training set was used as variables of consensus clustering. The result of consensus clustering indicated that 325 patients in the training set could be classified into two robust clusters with clustering stability increasing between k = 2 to k = 10 (Figure 1A–C; Supplementary Figure S1). Kaplan–Meier survival analysis showed that patients with gliomas in cluster2 ($n = 201$) had a significantly poorer prognosis than in cluster1 ($n = 124$; median OS: 555 vs. 1082 days; Figure 1D). Furthermore, differences in clinicopathological features between these two clusters were also found through Student's *t*-test and chi-square test (Supplementary Table S1). Cluster2 had a strong correlation with older age at diagnosis (median age: 46, $p < 0.001$), classical or mesenchymal subtypes (62.19%, $p < 0.001$), glioblastoma phenotype (58.21%, $p < 0.001$), IDH wildtype (70.65%, $p < 0.001$), and 1p/19q non-codeletion (82.59%, $p < 0.001$). By contrast, cluster1 mainly represented younger age at diagnosis (median age: 38, $p < 0.001$), proneural or neural subtypes (86.29%, $p < 0.001$), lower grade phenotype (78.23%, $p < 0.001$), and IDH mutation (87.09%, $p < 0.001$). Our results indicated that cellular response to IL-4-related gene set was involved in the malignancy of gliomas and strongly correlated to prognosis.

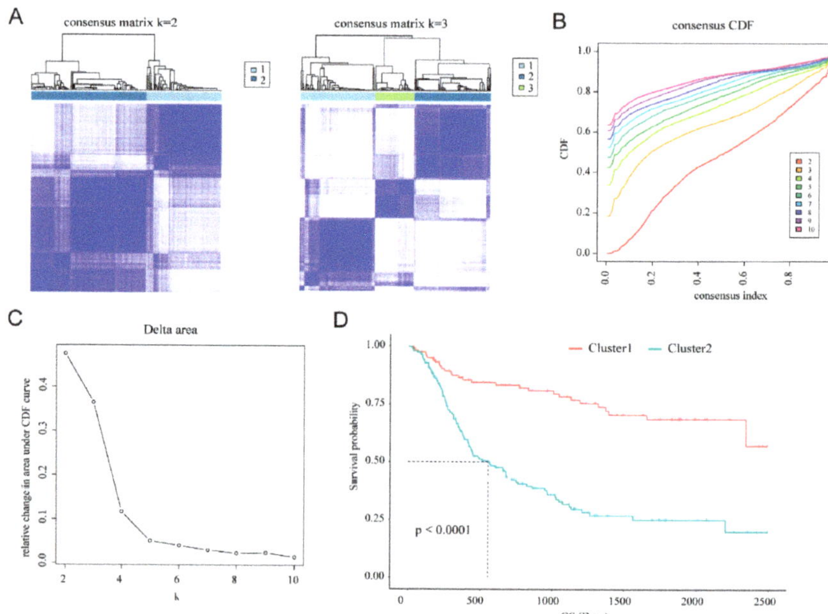

Figure 1. Classification of gliomas based on the cellular response to IL-4-related gene set in CGGA dataset. (**A**) Consensus clustering matrix of 325 CGGA samples for k = 2 and k = 3. (**B**) Consensus clustering CDF for k = 2 to k = 10. (**C**) Relative change in area under CDF curve for k = 2 to k = 10. (**D**) Survival analysis using Kaplan–Meier method for two clusters.

3.2. Identification of a 10-Gene Risk Signature Associated with Cellular Response to IL-4

Through univariate Cox regression analysis, all cellular responses to IL-4-related genes with high prognostic values ($p < 0.05$) were selected for further analysis in the CGGA training set. To identify the risk signature associated with cellular response to IL-4, genes with high prognostic values further underwent the LASSO regression analysis. After 10-fold cross validation, LASSO regression analysis generated 10 genes (CORO1A, FASN, HSPA5, IL2RG, LEF1, MCM2, NFIL3, PML, RPL3, TUBA1B) in total as active covariates to calculate the risk score (Figure 2; Table 1). The signature risk score of each patient in the training set and validating set was then calculated with the LASSO regression coefficients and expression value of these 10 genes through equations.

Table 1. Univariate Cox regression analysis and LASSO regression coefficients of 10 genes generated by LASSO regression analysis.

Gene	LASSO Regression Coefficient
CORO1A	0.020692273
FASN	−0.019673763
HSPA5	0.000533637
IL2RG	0.051870857
LEF1	0.038347165
MCM2	0.059818862
NFIL3	0.016612283
PML	0.101586488
RPL3	−0.002205126
TUBA1B	0.000593799

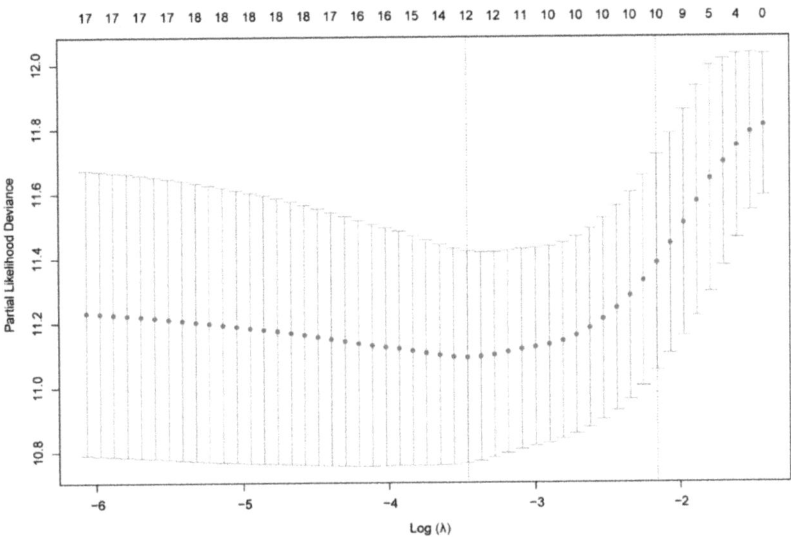

Figure 2. Lasso regression analysis of IL-4-associated genes with high prognostic values (Uni-Cox $p < 0.05$) generated 10 genes as covariates to calculate the gene risk signature. Dotted line indicated the most preferred λ value as the L1-regularization of Lasso regression. Box lines represented the variances of partial likelihood deviance of the λ value.

3.3. Cellular Response to IL-4-Related Gene Risk Signature Distinguished the Clinicopatho Logical Features of Gliomas

After calculating the 10-gene risk signature score of each patient, we observed that higher risk scores were found in glioblastoma than lower grade gliomas ($p < 0.001$), in classical and mesenchymal subtypes than other subtypes ($p < 0.001$), and in the IDH wildtype than the IDH mutation ($p < 0.001$) in the CGGA dataset (Figure 3A,C,E). A similar distributional pattern of the risk score was also observed in the TCGA dataset (Figure 3B,D,F). Then, we classified the patients in the training set into high-risk group and low-risk group by using median signature risk score as the cutoff value. Patients in the high-risk group were linked to older age at diagnosis (median age: 46.5, $p < 0.001$), classical or mesenchymal subtypes (79.01%, $p < 0.001$), glioblastoma phenotype (70.98%, $p < 0.001$), IDH wildtype (78.39%, $p < 0.001$), and 1p/19q non-codeletion (95.23%, $p < 0.001$, Supplementary Table S2). By contrast, patients in the low-risk group were associated with younger age at diagnosis (median age: 39, $p < 0.001$), proneural or neural subtypes (91.41%, $p < 0.001$), lower grade phenotype (82.21%, $p < 0.001$), and IDH mutation (80.98%, $p < 0.001$; Supplementary Table S2). In the TCGA dataset, we also observed that patients in the high-risk group were correlated with older age at diagnosis (median age: 54, $p < 0.001$), classical or mesenchymal subtypes (67.91%, $p < 0.001$), IDH wildtype (69.66%, $p < 0.001$), and 1p/19q non-codeletion (96.88%, $p < 0.001$, Supplementary Table S2), while patients in the low-risk group had a strong correlation with younger age at diagnosis (median age: 39, $p < 0.001$), proneural or neural subtypes (99.62%, $p < 0.001$), lower grade phenotype (99.70%, $p < 0.001$), and IDH mutation (97.03%, $p < 0.001$; Supplementary Table S2). These results indicated that the 10-gene risk signature associated with cellular response to IL-4 could distinguish the malignancy of gliomas.

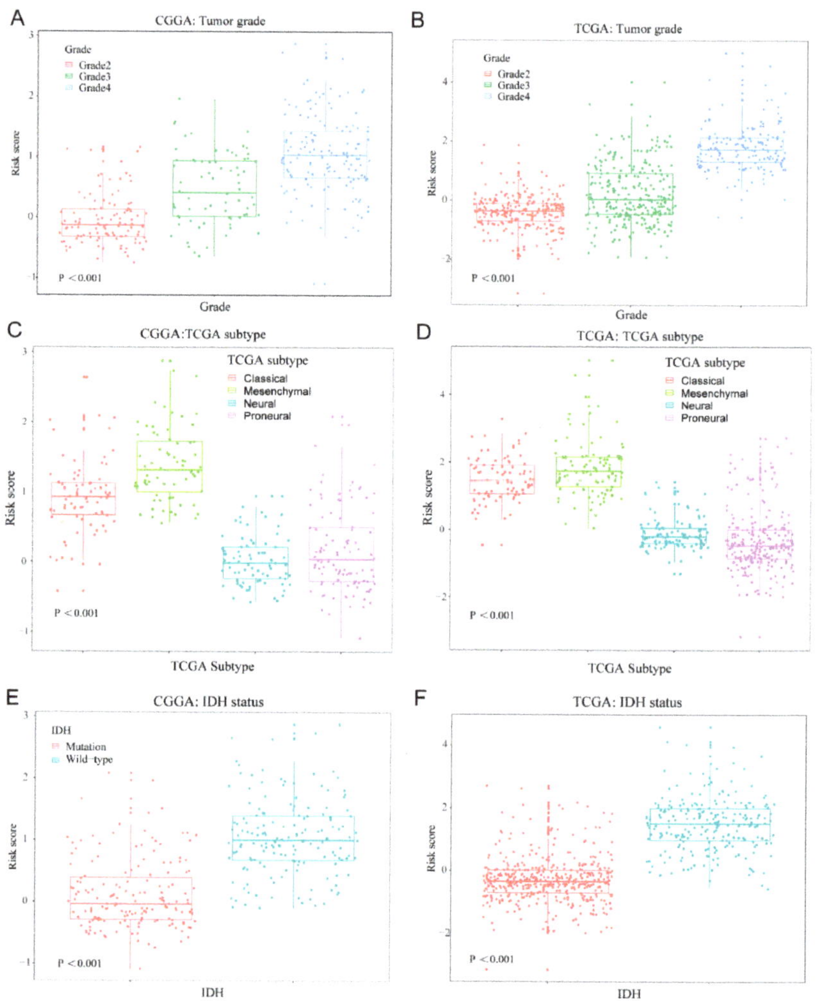

Figure 3. The 10-gene risk signature distinguished the clinicopathological features of gliomas. Distribution of the 8-gene risk signature with different tumor grades (**A**,**B**), TCGA subtypes (**C**,**D**) and IDH status (**E**,**F**).

3.4. Prognostic Value of Cellular Response to IL-4-Related Gene Risk Signature in All Grade Gliomas and Glioblastoma

In the CGGA dataset, Kaplan–Meier survival analysis revealed that patients in the high-risk group (n = 155) had a significantly poorer prognosis compared with patients in the low-risk group (n = 156; median OS: 376 days vs. NA; p < 0.001; Figure 4A). In the TCGA dataset, patients in the high-risk group (n = 327) were also found to have much shorter overall survival times than patients in the low-risk group (n = 338, median OS: 592 vs. 3200 days; p < 0.001; Figure 4C). After taking important clinical and molecular factors (including age, gender, WHO grade, IDH status, chemotherapy and radiotherapy) into account, univariate and multivariate Cox analysis further demonstrated that this risk score was an independent prognostic factor of prognosis in the CGGA dataset (Table 2). Cox proportional hazard model also found risk score could serve as an independent prognostic factor in the TCGA dataset (Table 2). When focusing on the GBM phenotype, we also observed that patients in the high-risk group (n = 71) had a shorter OS than patients in the low-risk group (n = 67) of the GBM phenotype in the CGGA dataset (median OS: 315 vs. 447 days; p = 0.0075; Figure 4B). Results in the TCGA dataset further validated the prognostic value of the risk signature in the GBM phenotype (median OS: 360 vs. 505 days; p = 0.0025; Figure 4D). These results indicated that our 10-gene risk signature associated with cellular response to IL-4 had high prognostic value in both all grade gliomas and glioblastoma.

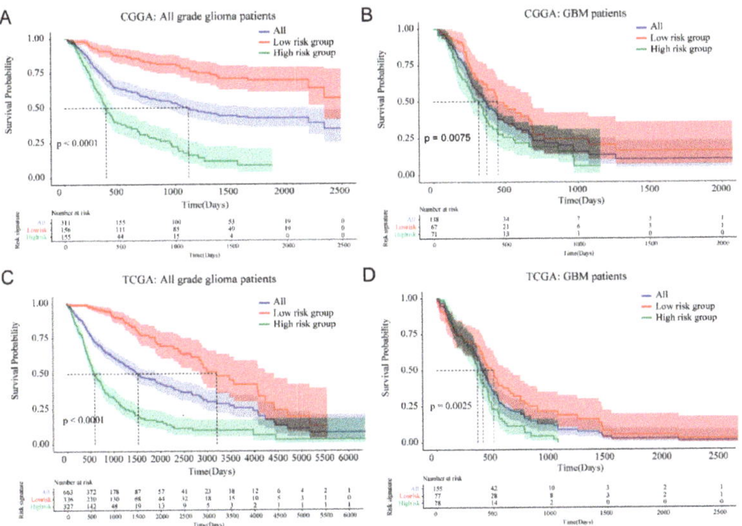

Figure 4. Prognostic value of 10-gene risk signature in CGGA and TCGA dataset. (**A,B**) Kaplan–Meier Survival curves for all grade gliomas and GBM patients in CGGA dataset. (**C,D**) Kaplan–Meier Survival curves for all grade gliomas and GBM patients in TCGA dataset.

Table 2. Univariate and multivariate Cox regression analysis of the clinical features and risk score for OS in CGGA and TCGA datasets.

Variables	Univariate Analysis			Multivariate Analysis		
	Hazard Ratio	**95% CI**	**p Value**	**Hazard Ratio**	**95% CI**	**p Value**
Training set CGGA RNA-seq cohort (n = 325)						
Age	1.0	1.0~1.1	<0.0001	1.01	0.99~1.03	0.22
Gender	1.2	0.83~1.7	0.37	1.32	0.88~1.96	0.176
Grade	5.9	4.1~8.6	<0.0001	1.79	1.07~3.0	0.026
IDH status	4.3	3~6.2	<0.0001	1.25	0.72~2.14	0.428
MGMT status	1.4	0.99~2.0	0.058	1.0	0.68~1.47	0.999
Chemotherapy	1.2	0.87~1.7	0.23	0.80	0.54~1.19	0.276
Radiotherapy	0.41	0.28~0.58	<0.0001	0.41	0.27~0.61	<0.001
Risk score	3.9	3.1~4.8	<0.0001	2.7	1.93~3.78	<0.001
Validation set TCGA RNA-seq cohort (n = 667)						
Age	1.1	1.1~1.1	<0.0001	1.03	1.02~1.04	<0.0001
Gender	1.2	0.96~1.6	0.11	1.48	1.06~2.1	0.021
Grade	9.1	6.9~12.0	<0.0001	1.63	1.04~2.6	0.033
IDH status	9.8	7.4~13.0	<0.0001	2.78	1.61~4.8	<0.001
MGMT status	3.3	2.5~4.3	<0.0001	1.19	0.81~1.7	0.367
Chemotherapy	0.41	0.27~0.61	<0.0001	0.63	0.4~1.0	0.052
Radiotherapy	2.1	1.5~2.9	<0.0001	1.05	0.63~1.7	0.843
Risk score	2.3	2.1~2.5	<0.0001	1.34	1.12~1.6	<0.033

3.5. Cellular Response to IL-4-Related Gene Risk Signature Was Correlated with Inhibited Immune Phenotype of Gliomas

After confirming the clustering and prognostic value of our IL-4-related gene risk signature, we then investigated the potential role of our risk signature in the immune phenotype of gliomas. We firstly calculated the abundances of 22 immune cell types in both the CGGA and TCGA datasets through the CIBERSORT package, and then presented the correlation between immune cells and our risk signature through heatmaps. In the CGGA datasets, higher risk score was strongly associated with less NK cells, less monocytes, less mast cells, more Tregs and more M0 macrophages (Figure 5). A similar phenotype was seen in the TCGA datasets, with higher risk score associated with less NK cells, less monocytes, less mast cells, more Tregs and more M0 macrophages (Supplementary Figure S2). With Pearson's test, our risk score was found to be strongly correlated with immunosuppressive factors (Figure 6), including CD274 (PD-L1), PDCD1 (PD-1), CTLA4, LAG3, HAVCR2 (TIM3), and so on. These results indicated that our cellular response to IL-4-related risk signature might be correlated with the inhibited immune phenotype of gliomas.

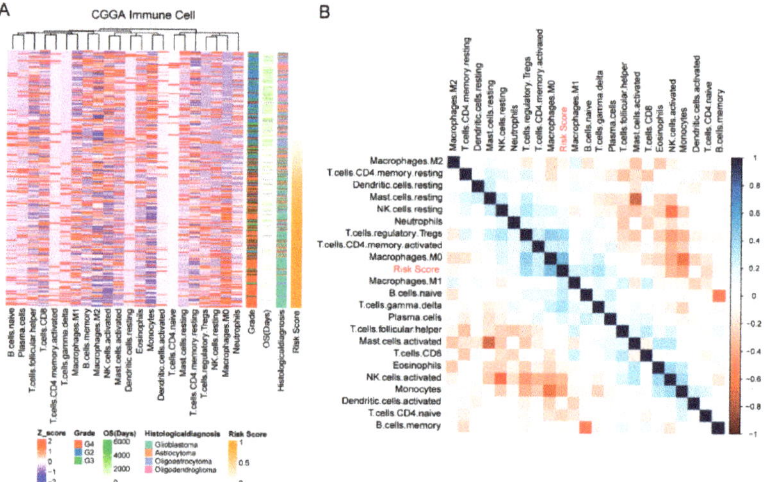

Figure 5. The 10-gene risk signature distinguished different local immune states in gliomas. (**A**) Analysis of heat map in CGGA dataset. (**B**) Correlation analysis in CGGA dataset.

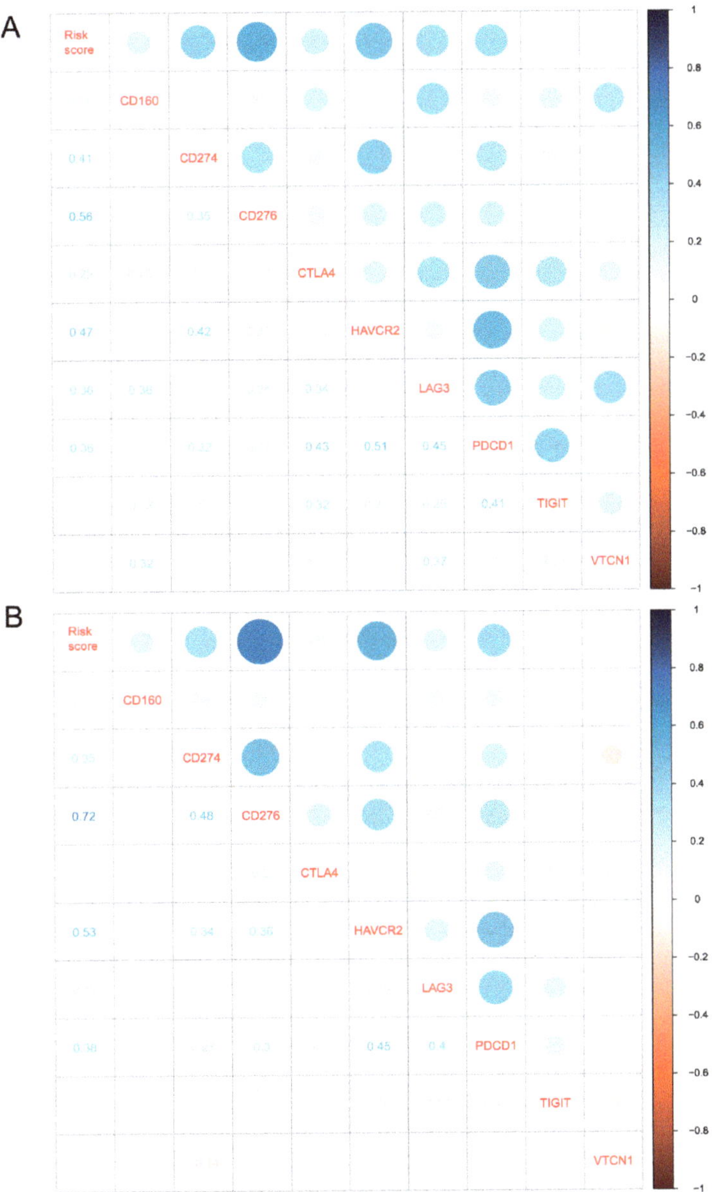

Figure 6. The 10-gene risk signature was strongly correlated with inhibited immune phenotype of gliomas. (**A**) Correlation analysis between risk score and immune suppressor in CGGA dataset. (**B**) Correlation analysis between risk score and immune suppressor in TCGA dataset.

4. Discussion

Gliomas are the most frequent tumors of the central nervous system, with extremely poor prognosis in adults. In 2016, the WHO classification of CNS tumors was revised for the first time using molecular and genetic biomarkers (IDH mutation and 1p/19q codeletion) to classify gliomas [24]. Then, the new WHO classification published in 2021 included more molecular biomarkers [25]. Since then, a diverse set of biomarkers have been implicated as prognostic indicators in gliomas, including checkpoint molecules (PD-1, CTLA-4, TIM-3) [26–28], growth/angiogenesis proteins (EGFR) [29], and cytokines (TGF-β, IL-4, IL-13) [30,31]. Among them, the effects of IL-4/IL-4 receptors were investigated in the last decade [32]. Some research advocated IL-4/IL-4 receptors could serve as excellent biomarkers and immunotherapeutic targets [33], while others found polymorphisms in IL-4 genes were not associated with glioma risk independently [32].

With the development of bioinformatic technologies, gene risk signature analysis emerged as a useful method to identify prognostic signatures in almost all kinds of cancers [34]. In gliomas, gene risk signature analysis also identified several new biomarkers including immune-related, metabolism-related, and inflammation-related gene signatures [15,35–37]. Here, we introduced a cellular response to IL-4-related gene signature as a newly discovered biomarker of clinicopathological features, prognosis, and immune phenotype of gliomas. We firstly confirmed that 28 genes on the cellular response to IL-4 pathway had the ability to distinguish the key clinicopathological features of gliomas in both CGGA and TCGA datasets. Then, we built the cellular response to IL-4 gene risk signature through LASSO regression analysis. Our cellular response to IL-4 gene risk signature was found to be strongly correlated with previously confirmed clinical features of gliomas including WHO grade, molecular subtypes, 1p/19q codeletion status and IDH status [24]. Using our gene risk signature, patients with higher risk score tend to be associated with the higher WHO glioma grade, the more invasive molecular subtypes (classical and mesenchymal), 1p/19q non-codeletion and IDH wide type, which represented worse prognosis. By contrast, the lower WHO glioma grade, the less invasive TCGA subtypes (proneural and neural subtype) and IDH mutation were preferentially associated with patients in the lower risk group. Our IL-4-related signature can work as a molecular biomarker to classify glioma patients combined with current genetic biomarkers, which will be beneficial to predict survival more precisely and even can be used to predict clinical response to adjuvant therapies such as immunotherapy. By analyzing the Kaplan–Meier survival curve and Cox proportional hazards model, we found our gene risk signature could serve as an independent prognostic marker of gliomas. In both CGGA and TCGA datasets, patients in the high-risk group had a significantly poorer prognosis compared with patients in the low-risk group (CGGA, median OS: 376 days vs. NA; TCGA, median OS: 592 vs. 3200 days). Considering the GBM phenotype, patients in the high-risk group also had a significantly shorter OS than patients in the low-risk group (CGGA, median OS: 315 vs. 447 days; TCGA, median OS: 360 vs. 505 days).

Next, we focus on the relationship between our gene risk signature and immune phenotype of the gliomas by CIBERSORT algorithm. In addition to the traditional surgery, chemotherapy and radiotherapy, immunotherapy has been regarded as the next generation approach in the treatment of gliomas. A range of different immunotherapies such as PD-1/PD-L1 checkpoint inhibitors, chimeric antigen receptor-T cells (CAR-T) and adoptive T cell strategies have currently been actively investigated in patients with gliomas [4–6]. Unfortunately, negative outcomes of these clinical trials challenge the concept of immunotherapy as a single modality treatment of gliomas [38]. Local immunosuppression in the glioma microenvironment might be responsible for the failure of current immunotherapies. The pathologic findings of most patients with gliomas showed a typical 'cold' tumor microenvironment with exhausted CD8+T cells and enriched Tregs [38,39]. Our cellular response to IL-4 gene risk signature showed to be highly correlated with Tregs, and higher risk scores were associated with more infiltrating Tregs. However, our risk signature was not found to be correlated with the infiltrating CD8+T cells. In addition to the adoptive immune cells, innate immune cells such as macrophages and NK cells have been a new trend in the research of glioma immunotherapy [40]. We also calculated the correlation between our risk signature and innate immune cells and found higher risk scores were strongly associated with more M0 macrophages and less NK cells. Unlike the classically activated immunostimulant M1 phenotype macrophages and alternatively activated immunosuppressive M2 phenotype, M0 macrophages are nonpolarized macrophages and found to be positively associated with malignant phenotypes of gliomas [41]. Contrary to the traditional concept, glioma-infiltrated macrophages were recently found to resemble the M0 macrophage phenotype instead of M1 or M2 phenotype [42]. The genetic and molecular mechanism of immunosuppressive M0 macrophages in gliomas remains unclear. According to our bioinformatic analysis and current dogma that Tregs secrete IL-4 to trigger the development of tumor-associated macrophages with immunosuppressive properties [43], we considered our cellular response to IL-4-related gene signature as evidence of the immunosuppressive mechanism of M0 macrophages in the glioma microenvironment. Moreover, NK cells are known as innate immune effective cells but are frequently exhausted in the glioma microenvironment [40]. Our IL-4-related risk signature might also indicate a potential genetic pathway of NK cell exhaustion in the gliomas.

5. Conclusions

In summary, we provided a cellular response to the IL-4-related gene signature as an excellent clinicopathological, prognostic and immune biomarker of gliomas in this study. In the future, the pathway involved in IL-4 to regulate the infiltration of immune cells will be an intriguing topic, especially in patients with high score, which may be beneficial to develop a novel immunotherapy or clinical biomarker. We believe that our bioinformatic analysis might provide new insight for understanding the IL-4 involved mechanism of gliomas.

Supplementary Materials: The following supporting information can be downloaded at: https://www.mdpi.com/article/10.3390/brainsci12020181/s1, Figure S1: Consensus clustering matrix of 325 CGGA samples for k = 4 to k = 10; Figure S2: The 10-gene risk signature distinguished different local immune states in gliomas. (A) Analysis of heat map in TCGA dataset. (B) Correlation analysis in TCGA dataset; Table S1: Clinicopathological features of two clusters classified by consensus clustering based on cellular response to IL-4-related gene set in CGGA dataset; Table S2: Clinicopathological features of two gene risk groups in CGGA and TCGA dataset.

Author Contributions: Conceptualization, Y.Q. and L.X.; Data curation, X.Y. and C.T.; Methodology, X.Y.; Visualization, C.J.; Writing—original draft, Y.Q.; Writing—review and editing, L.X. All authors were involved in writing the paper and had final approval of the submitted and published versions. All authors have read and agreed to the published version of the manuscript.

Funding: This study was funded by the Shanghai Anticancer Association EYAS PROJECT (SACA-CY20B09) and the National Natural Science Foundation of China (81772672).

Institutional Review Board Statement: Not applicable.

Informed Consent Statement: Informed consent was obtained from all subjects involved in the study.

Data Availability Statement: Publicly available datasets were used in this study. The datasets analyzed in the present study were retrieved from The Cancer Genome Atlas and Chinese Glioma Genome Atlas databases.

Acknowledgments: We would like to acknowledge the reviewers for their helpful comments on this paper.

Conflicts of Interest: The authors declare no conflict of interest.

References

1. Montemurro, N.; Fanelli, G.N.; Scatena, C.; Ortenzi, V.; Pasqualetti, F.; Mazzanti, C.M.; Morganti, R.; Paiar, F.; Naccarato, A.G.; Perrini, P. Surgical outcome and molecular pattern characterization of recurrent glioblastoma multiforme: A single-center retrospective series. *Clin. Neurol. Neurosurg.* **2021**, *207*, 106735. [CrossRef] [PubMed]
2. Kazmi, F.; Soon, Y.Y.; Leong, Y.H.; Koh, W.Y.; Vellayappan, B. Re-irradiation for recurrent glioblastoma (GBM): A systematic review and meta-analysis. *J. Neuro-Oncol.* **2019**, *142*, 79–90. [CrossRef] [PubMed]
3. Jackson, C.M.; Choi, J.; Lim, M. Mechanisms of immunotherapy resistance: Lessons from glioblastoma. *Nat. Immunol.* **2019**, *20*, 1100–1109. [CrossRef] [PubMed]

4. de Groot, J.; Penas-Prado, M.; Alfaro-Munoz, K.; Hunter, K.; Pei, B.L.; O'Brien, B.; Weathers, S.P.; Loghin, M.; Kamiya Matsouka, C.; Yung, W.K.A.; et al. Window-of-opportunity clinical trial of pembrolizumab in patients with recurrent glioblastoma reveals predominance of immune-suppressive macrophages. *Neuro Oncol.* **2020**, *22*, 539–549. [CrossRef] [PubMed]

5. Bagley, S.J.; Desai, A.S.; Linette, G.P.; June, C.H.; O'Rourke, D.M. CAR T-cell therapy for glioblastoma: Recent clinical advances and future challenges. *Neuro Oncol.* **2018**, *20*, 1429–1438. [CrossRef] [PubMed]

6. Choi, B.D.; Maus, M.V.; June, C.H.; Sampson, J.H. Immunotherapy for Glioblastoma: Adoptive T-cell Strategies. *Clin. Cancer Res.* **2019**, *25*, 2042–2048. [CrossRef]

7. Buonfiglioli, A.; Hambardzumyan, D. Macrophages and microglia: The cerberus of glioblastoma. *Acta Neuropathol. Commun.* **2021**, *9*, 54. [CrossRef] [PubMed]

8. Lupo, K.B.; Matosevic, S. CD155 immunoregulation as a target for natural killer cell immunotherapy in glioblastoma. *J. Hematol. Oncol.* **2020**, *13*, 76. [CrossRef] [PubMed]

9. Daubon, T.; Hemadou, A.; Romero Garmendia, I.; Saleh, M. Glioblastoma Immune Landscape and the Potential of New Immunotherapies. *Front. Immunol.* **2020**, *11*, 585616. [CrossRef]

10. LaPorte, S.L.; Juo, Z.S.; Vaclavikova, J.; Colf, L.A.; Qi, X.; Heller, N.M.; Keegan, A.D.; Garcia, K.C. Molecular and structural basis of cytokine receptor pleiotropy in the interleukin-4/13 system. *Cell* **2008**, *132*, 259–272. [CrossRef]

11. Joshi, B.H.; Leland, P.; Lababidi, S.; Varrichio, F.; Puri, R.K. Interleukin-4 receptor alpha overexpression in human bladder cancer correlates with the pathological grade and stage of the disease. *Cancer Med.* **2014**, *3*, 1615–1628. [CrossRef] [PubMed]

12. Todaro, M.; Lombardo, Y.; Francipane, M.G.; Alea, M.P.; Cammareri, P.; Iovino, F.; Di Stefano, A.B.; Di Bernardo, C.; Agrusa, A.; Condorelli, G.; et al. Apoptosis resistance in epithelial tumors is mediated by tumor-cell-derived interleukin-4. *Cell Death Differ.* **2008**, *15*, 762–772. [CrossRef] [PubMed]

13. Setrerrahmane, S.; Xu, H. Tumor-related interleukins: Old validated targets for new anti-cancer drug development. *Mol. Cancer* **2017**, *16*, 153. [CrossRef] [PubMed]

14. Scheurer, M.E.; Amirian, E.; Cao, Y.; Gilbert, M.R.; Aldape, K.D.; Kornguth, D.G.; El-Zein, R.; Bondy, M.L. Polymorphisms in the interleukin-4 receptor gene are associated with better survival in patients with glioblastoma. *Clin. Cancer Res.* **2008**, *14*, 6640–6646. [CrossRef] [PubMed]

15. Liu, Y.-Q.; Chai, R.-C.; Wang, Y.-Z.; Wang, Z.; Liu, X.; Wu, F.; Jiang, T. Amino acid metabolism-related gene expression-based risk signature can better predict overall survival for glioma. *Cancer Sci.* **2019**, *110*, 321–333. [CrossRef] [PubMed]

16. Zhao, S.; Cai, J.; Li, J.; Bao, G.; Li, D.; Li, Y.; Zhai, X.; Jiang, C.; Fan, L. Bioinformatic Profiling Identifies a Glucose-Related Risk Signature for the Malignancy of Glioma and the Survival of Patients. *Mol. Neurobiol.* **2017**, *54*, 8203–8210. [CrossRef]

17. Aoki, K.; Nakamura, H.; Suzuki, H.; Matsuo, K.; Kataoka, K.; Shimamura, T.; Motomura, K.; Ohka, F.; Shiina, S.; Yamamoto, T.; et al. Prognostic relevance of genetic alterations in diffuse lower-grade gliomas. *Neuro-Oncology* **2018**, *20*, 66–77. [CrossRef]

18. Ceccarelli, M.; Barthel, F.P.; Malta, T.M.; Sabedot, T.S.; Salama, S.R.; Murray, B.A.; Morozova, O.; Newton, Y.; Radenbaugh, A.; Pagnotta, S.M.; et al. Molecular Profiling Reveals Biologically Discrete Subsets and Pathways of Progression in Diffuse Glioma. *Cell* **2016**, *164*, 550–563. [CrossRef]

19. Wilkerson, M.D.; Hayes, D.N. ConsensusClusterPlus: A class discovery tool with confidence assessments and item tracking. *Bioinformatics* **2010**, *26*, 1572–1573. [CrossRef]

20. Tibshirani, R. Regression Shrinkage and Selection Via the Lasso. *J. R. Stat. Soc. Ser. B* **1996**, *58*, 267–288. [CrossRef]

21. Friedman, J.H.; Hastie, T.; Tibshirani, R. Regularization Paths for Generalized Linear Models via Coordinate Descent. *J. Stat. Softw.* **2010**, *33*, 22. [CrossRef]

22. Newman, A.M.; Liu, C.L.; Green, M.R.; Gentles, A.J.; Feng, W.; Xu, Y.; Hoang, C.D.; Diehn, M.; Alizadeh, A.A. Robust enumeration of cell subsets from tissue expression profiles. *Nat. Methods* **2015**, *12*, 453–457. [CrossRef] [PubMed]

23. Gu, Z.; Eils, R.; Schlesner, M. Complex heatmaps reveal patterns and correlations in multidimensional genomic data. *Bioinformatics* **2016**, *32*, 2847–2849. [CrossRef] [PubMed]

24. Wesseling, P.; Capper, D. WHO 2016 Classification of gliomas. *Neuropathol. Appl. Neurobiol.* **2018**, *44*, 139–150. [CrossRef]

25. Louis, D.N.; Perry, A.; Wesseling, P.; Brat, D.J.; Cree, I.A.; Figarella-Branger, D.; Hawkins, C.; Ng, H.K.; Pfister, S.M.; Reifenberger, G.; et al. The 2021 WHO Classification of Tumors of the Central Nervous System: A summary. *Neuro-Oncol.* **2021**, *23*, 1231–1251. [CrossRef] [PubMed]

26. Cloughesy, T.F.; Mochizuki, A.Y.; Orpilla, J.R.; Hugo, W.; Lee, A.H.; Davidson, T.B.; Wang, A.C.; Ellingson, B.M.; Rytlewski, J.A.; Sanders, C.M.; et al. Neoadjuvant anti-PD-1 immunotherapy promotes a survival benefit with intratumoral and systemic immune responses in recurrent glioblastoma. *Nat. Med.* **2019**, *25*, 477–486. [CrossRef] [PubMed]

27. Liu, F.; Huang, J.; Liu, X.; Cheng, Q.; Luo, C.; Liu, Z. CTLA-4 correlates with immune and clinical characteristics of glioma. *Cancer Cell Int.* **2020**, *20*, 7. [CrossRef] [PubMed]

28. Kim, J.E.; Patel, M.A.; Mangraviti, A.; Kim, E.S.; Theodros, D.; Velarde, E.; Liu, A.; Sankey, E.W.; Tam, A.; Xu, H.; et al. Combination Therapy with Anti-PD-1, Anti-TIM-3, and Focal Radiation Results in Regression of Murine Gliomas. *Clin. Cancer Res.* **2017**, *23*, 124–136. [CrossRef]

29. Saadeh, F.S.; Mahfouz, R.; Assi, H.I. EGFR as a clinical marker in glioblastomas and other gliomas. *Int. J. Biol. Markers* **2018**, *33*, 22–32. [CrossRef]

30. Tang, J.; Yu, B.; Li, Y.; Zhang, W.; Alvarez, A.A.; Hu, B.; Cheng, S.Y.; Feng, H. TGF-β-activated lncRNA LINC00115 is a critical regulator of glioma stem-like cell tumorigenicity. *EMBO Rep.* **2019**, *20*, e48170. [CrossRef]

31. Chen, P.; Chen, C.; Chen, K.; Xu, T.; Luo, C. Polymorphisms in IL-4/IL-13 pathway genes and glioma risk: An updated meta-analysis. *Tumour. Biol.* **2015**, *36*, 121–127. [CrossRef] [PubMed]
32. Shimamura, T.; Husain, S.R.; Puri, R.K. The IL-4 and IL-13 pseudomonas exotoxins: New hope for brain tumor therapy. *Neurosurg. Focus* **2006**, *20*, E11. [CrossRef]
33. Hung, A.L.; Garzon-Muvdi, T.; Lim, M. Biomarkers and Immunotherapeutic Targets in Glioblastoma. *World Neurosurg.* **2017**, *102*, 494–506. [CrossRef]
34. Zhang, L.; Zhang, Z.; Yu, Z. Identification of a novel glycolysis-related gene signature for predicting metastasis and survival in patients with lung adenocarcinoma. *J. Transl. Med.* **2019**, *17*, 423. [CrossRef] [PubMed]
35. Zhang, M.; Wang, X.; Chen, X.; Zhang, Q.; Hong, J. Novel Immune-Related Gene Signature for Risk Stratification and Prognosis of Survival in Lower-Grade Glioma. *Front. Genet.* **2020**, *11*, 363. [CrossRef]
36. Qi, Y.; Chen, D.; Lu, Q.; Yao, Y.; Ji, C. Bioinformatic Profiling Identifies a Fatty Acid Metabolism-Related Gene Risk Signature for Malignancy, Prognosis, and Immune Phenotype of Glioma. *Dis. Markers* **2019**, *2019*, 3917040. [CrossRef]
37. Wang, Z.; Gao, L.; Guo, X.; Feng, C.; Lian, W.; Deng, K.; Xing, B. Development and validation of a nomogram with an autophagy-related gene signature for predicting survival in patients with glioblastoma. *Aging* **2019**, *11*, 12246–12269. [CrossRef]
38. Lim, M.; Xia, Y.; Bettegowda, C.; Weller, M. Current state of immunotherapy for glioblastoma. *Nat. Rev. Clin. Oncol.* **2018**, *15*, 422–442. [CrossRef] [PubMed]
39. Woroniecka, K.; Chongsathidkiet, P.; Rhodin, K.; Kemeny, H.; Dechant, C.; Farber, S.H.; Elsamadicy, A.A.; Cui, X.; Koyama, S.; Jackson, C.; et al. T-Cell Exhaustion Signatures Vary with Tumor Type and Are Severe in Glioblastoma. *Clin. Cancer Res. Off. J. Am. Assoc. Cancer Res.* **2018**, *24*, 4175–4186. [CrossRef] [PubMed]
40. Friebel, E.; Kapolou, K.; Unger, S.; Núñez, N.G.; Utz, S.; Rushing, E.J.; Regli, L.; Weller, M.; Greter, M.; Tugues, S.; et al. Single-Cell Mapping of Human Brain Cancer Reveals Tumor-Specific Instruction of Tissue-Invading Leukocytes. *Cell* **2020**, *181*, 1626–1642.e1620. [CrossRef]
41. Huang, L.; Wang, Z.; Chang, Y.; Wang, K.; Kang, X.; Huang, R.; Zhang, Y.; Chen, J.; Zeng, F.; Wu, F.; et al. EFEMP2 indicates assembly of M0 macrophage and more malignant phenotypes of glioma. *Aging* **2020**, *12*, 8397–8412. [CrossRef] [PubMed]
42. Gabrusiewicz, K.; Rodriguez, B.; Wei, J.; Hashimoto, Y.; Healy, L.M.; Maiti, S.N.; Thomas, G.; Zhou, S.; Wang, Q.; Elakkad, A.; et al. Glioblastoma-infiltrated innate immune cells resemble M0 macrophage phenotype. *JCI Insight.* **2016**, *1*, e85841. [CrossRef] [PubMed]
43. Mantovani, A.; Marchesi, F.; Malesci, A.; Laghi, L.; Allavena, P. Tumour-associated macrophages as treatment targets in oncology. *Nat. Rev. Clin. Oncol.* **2017**, *14*, 399–416. [CrossRef] [PubMed]

MDPI

St. Alban-Anlage 66

4052 Basel

Switzerland

Tel. +41 61 683 77 34

Fax +41 61 302 89 18

www.mdpi.com

Brain Sciences Editorial Office

E-mail: brainsci@mdpi.com

www.mdpi.com/journal/brainsci